OPENING A WINDOW TO THE WEST

The Foreign Concession at Kōbe, Japan, 1868–1899

After more than two centuries of self-seclusion, Japan finally opened its borders to Western traders and influences in the 1850s. However, Westerners were restricted to a handful of Foreign Concessions set adjacent to selected Japanese cities, where they could fashion a working urban space suited to their own cultural patterns. This provided the Japanese with a microscopic lens on Western ways of behaviour and commerce. Kōbe was one of these treaty ports, and its Foreign Concession, along with that at Yokohama, became the most vibrant and successful of these settlements.

The first book-length study of Kōbe's Foreign Concession, *Opening a Window to the West* situates Kōbe within the larger pattern of globalization occurring throughout East Asia in the nineteenth century. Detailing the form and evolution of the settlement, its social and economic composition, and its specific mercantile trading features, this vivid microstudy illuminates the making of Kōbe during these critical decades of growth and development.

(Japan & Global Society)

PETER ENNALS is a professor emeritus of Geography and Environment at Mount Allison University. He is co-author of *Homeplace: The Making of the Canadian Dwelling over Three Centuries*, and has contributed to the *Historical Atlas of Canada* and the *Dictionary of Canadian Biography*.

JAPAN AND GLOBAL SOCIETY

Editors: Akira Iriye, *Harvard University*; Masato Kimura, *Shibusawa Eiichi Memorial Foundation*; David A. Welch, *Balsillie School of International Affairs, University of Waterloo*

How has Japan shaped, and been shaped by, globalization – politically, economically, socially, and culturally? How has its identity, and how have its objectives, changed? *Japan and Global Society* explores Japan's past, present, and future interactions with the Asia Pacific and the world from a wide variety of disciplinary and interdisciplinary perspectives and through diverse paradigmatic lenses. Titles in this series are intended to showcase international scholarship on Japan and its regional neighbours that will appeal to scholars in disciplines in both the humanities and the social sciences.

Japan and Global Society is supported by generous grants from the Shibusawa Eiichi Memorial Foundation and the University of Missouri – St Louis.

Editorial Advisory Board

Frederick R. Dickinson, University of Pennsylvania
Michael Donnelly, University of Toronto
Joel Glassman, University of Missouri – St Louis
Izumi Koide, Shibusawa Eiichi Memorial Foundation
Gil Latz, Portland State University
Michael A. Schneider, Knox College
Patricia G. Steinhoff, University of Hawaii at Manoa
Patricia Wetzel, Portland State University

For a list of books published in the series, see page 239.

PETER ENNALS

Opening a Window to the West

The Foreign Concession at Kōbe, Japan, 1868–1899

UNIVERSITY OF TORONTO PRESS
Toronto Buffalo London

© University of Toronto Press 2014
Toronto Buffalo London
www.utppublishing.com

ISBN 978-1-4426-4602-5 (cloth)
ISBN 978-1-4426-1416-1 (paper)

Library and Archives Canada Cataloguing in Publication

Ennals, Peter, author
Opening a window to the West : the foreign concession at Kōbe, Japan, 1868–1899 / Peter Ennals.

(Japan and global society)
Includes bibliographical references and index.
ISBN 978-1-4426-4602-5 (bound). – ISBN 978-1-4426-1416-1 (pbk.)

1. Concessions – Japan – Kōbe-shi – History – 19th century.
2. Land use – Japan – Kōbe-shi – History – 19th century.   3. Kōbe-shi (Japan) – Commerce – History – 19th century.   4. Kōbe-shi (Japan) – Historical geography.   5. Kōbe-shi (Japan) – Economic conditions – 19th century.   6. Kōbe-shi (Japan) – Social conditions – 19th century.   I. Title.   II. Series: Japan and global society

HF3830.K6E56 2013   338.0952'187409034   C2013-905576-2

University of Toronto Press acknowledges the financial assistance to its publishing program of the Canada Council for the Arts and the Ontario Arts Council.

 Canada Council for the Arts   Conseil des Arts du Canada

University of Toronto Press acknowledges the financial support of the Government of Canada through the Canada Book Fund for its publishing activities.

*Dedicated
to
Professor George Oshima
and to
Kwansei Gakuin University*

Contents

List of Illustrations ix

Abbreviations xi

Preface xv

Acknowledgments xxi

1 Setting the Stage: The Role of Ports in the Encounter between East and West in Japan 3

2 The Creation of Kōbe's Foreign Concession 22

3 Establishing Municipal Government and Services in the Concession 42

4 Forging an Economy: The Basis for Mercantile Trade 67

5 Finding a Mercantile Staple for Kōbe: The Tea and Silk Trades 88

6 The Morphology of the Settlement and the Development of a Pleasing Townscape 111

7 Life at the End of the World: Forming an Expatriate Society in Kōbe 146

8 Measuring Success in the Concession 171

Glossary 187

Explanatory Notes 189

Notes 191

Bibliography 215

Index 229

List of Illustrations

Maps

0.1 General reference map of Japan showing major cities and features noted in the text   xiii
0.2 Map of Kōbe and its hinterland during the treaty port era   xiv
1.1 Economic land use in early-nineteenth-century Calcutta   18
2.1 Map locating the Foreign Concession in its setting east of Hiōgo City, ca. 1870   24
2.2 Map of the Foreign Concessions at Hankow, China, ca. 1880, showing the separate national sectors   29
2.3 A map of Shanghai, ca. 1855, showing the British sector of the International Settlement, and particularly the gridiron pattern of the earliest area of settlement   31
2.4 Sale of lots in the Hiōgo Foreign Concession, by date of auction   35
6.1 Map of the Concession and surrounding areas, 1880   116
6.2 The shifting location of select businesses, 1868–88   122
6.3 A map of functional land use in and around the Foreign Concession, ca. 1880   141

Figures

1.1 Bowden's Mercantile Triangle model   20
2.1 Kyo-machi during the 1870s, looking northeast   40
3.1 Political structures relating to municipal governance   45
4.1 Ships entering the port of Kōbe, 1868–80, by number and nationality   78

4.2   Value of trade at Kōbe, 1873–99   86
6.1   Foreigner's houses preserved in Kōbe   128
6.2
& 6.3 The Foreign Settlement at Kōbe, 1878 (C.B. Bernard)   130
6.4   The Bund, from the corner of Naniwa-machi   133
6.5   Akashi-machi, ca. 1870, looking north from the Bund   134
6.6   The Walsh Hall & Co. premises, Lot 2, the Bund, Hiōgo   134
6.7   Motomachi Street in the Native Town   135
6.8   The "Native Bund" at Hiōgo, showing the Hiōgo Hotel   135
6.9   A view from within a compound in the Concession   136
6.10 Looking northward from the Concession, ca. 1880   137
6.11 A spatial model of Kōbe as a mercantile settlement   143

Tables

4.1 Trade at the port of Kōbe, 1868   81
4.2 Summary of trade through the ports of Kōbe and Ōsaka, 1870   84
4.3 Imports and exports at the port of Kōbe, in yen, 1873–99   85
4.4 Economic fluctuations in Japan, 1868–97 with reference to broader external business cycles   86
5.1 Movement of silk from the ports of Hiōgo and Ōsaka, 1868   95
5.2 Tea exports from the foreign port of Hiōgo, 1869–83   99
5.3 Shipments of tea by firm at Kōbe, 1879 and 1880   106
5.4 Destination of annual exports from Ōsaka and Hiōgo, 1 May 1874–30 April 1884   109
5.5 Destinations of tea shipped from the port of Hiōgo in 1894   110
7.1 Foreign population of Hiōgo by nationality, 1870–96   148
7.2 Women and children in the foreign population at Hiōgo, 1878–93   150
7.3 Foreigners' businesses and occupations, 1871   163

Abbreviations

FO [British] Foreign Office
NARA [US] National Archives and Records Administration
PRO [British] Public Record Office

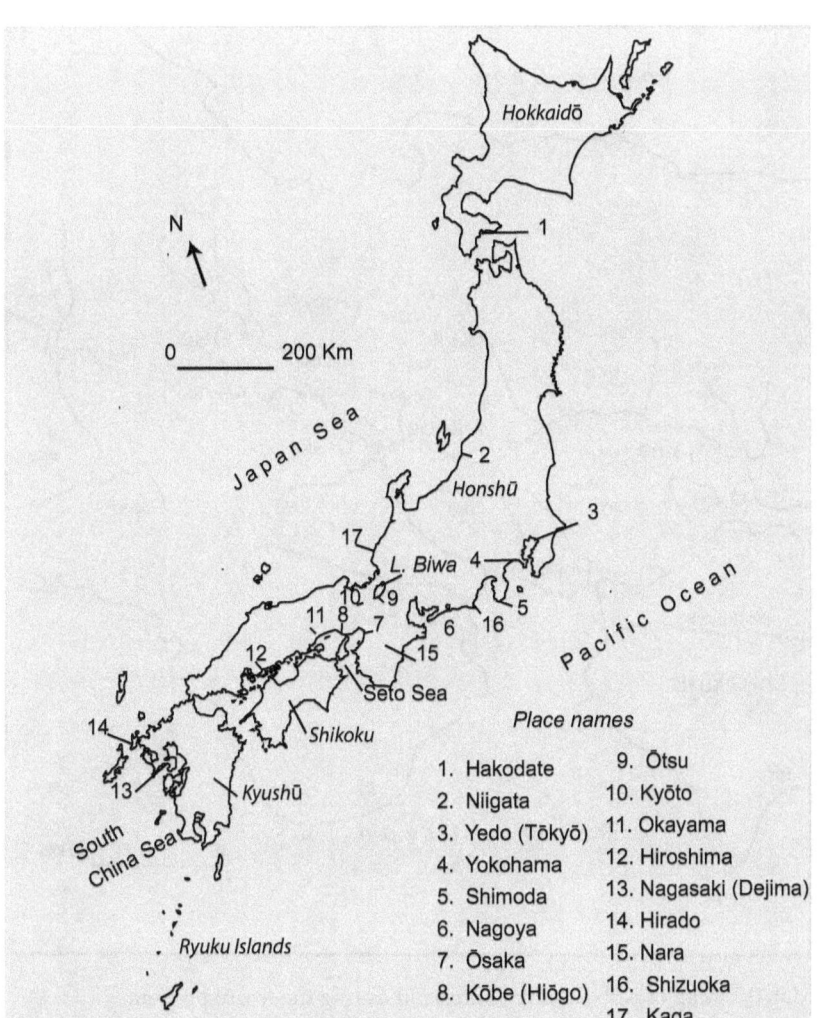

Map 0.1 General reference map of Japan showing major cities and features noted in the text

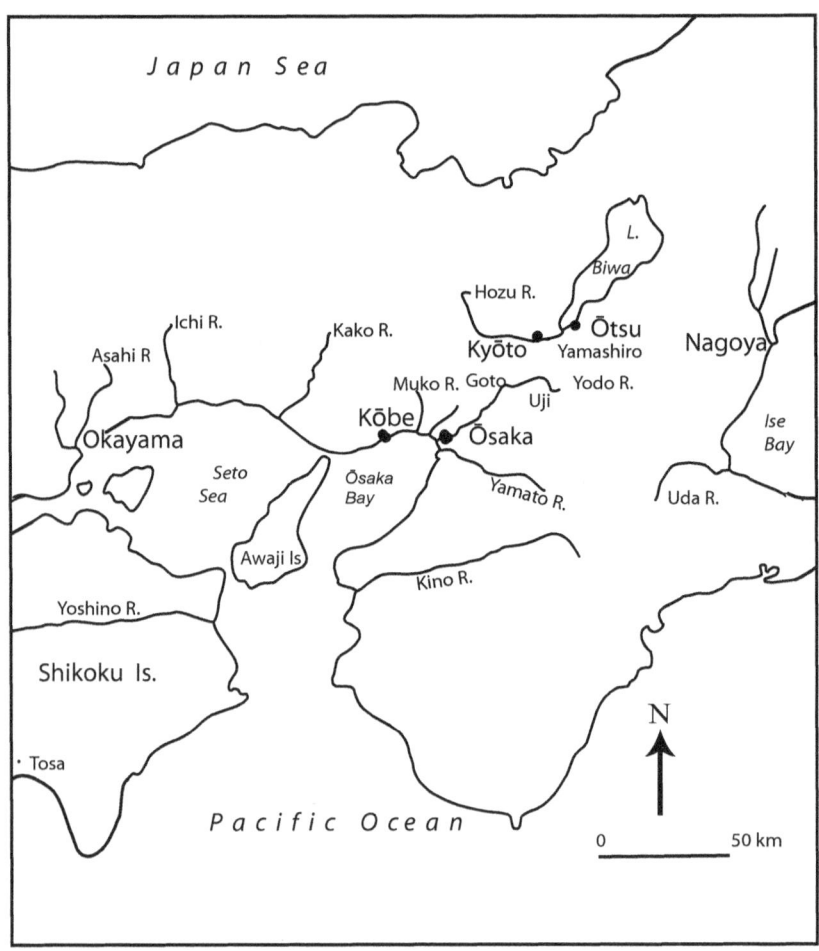

Map 0.2 Map of Kōbe and its hinterland during the treaty port era

Preface

[Kōbe] is packed along the shore for a distance of three miles [4.8 km], the Kioto [sic] and Hiogo Railway ... running through the town down to a pier which enables ships of large tonnage to receive cargo direct from the railway trucks. The Foreign Concession, beautifully and regularly laid out on a grand scale for the population which [it] has never attracted, is at the east end. It is a "model settlement," well lighted with gas, and supplied with water, kept methodically clean, and efficiently cared for by the police. The Bund has a stone embankment, a grass parade, and a magnificent carriage-road, with the British, American and German Consulate, and some imposing foreign residences on the other side. Several short streets run back from this, crossing a long one parallel with the Bund. The side-walks are very broad, and well paved with stones laid edgewise, with curb-stones and handsome paved waterways, and the carriage-roads are broad and beautifully kept. The foreign houses are spacious and solid, and the railroad, and the station and its environments, are of the most approved English construction.

But where are the people? Roads without houses, carriage-ways without carriages, side-walks without foot-passengers, and a solitude so dreary that three men stopping in the street to talk is a sight which might collect the rest of the community to stare at it, are features of what was intended to be an important place. It is mainly English, but there are only about 170 residents, and this includes all the British firms from Osaka, who migrated here when the railroad was opened.

... A number of foreign wooden houses straggle up the foot-hills at the back, some of them unmistakable English bungalows, while those which look like Massachusetts homesteads are occupied by American missionaries. In spite of the solitude and stagnation of the streets of the settlement, Kobe is a pertinaciously cheerful-looking place.

... A large native town has grown up at Kobe, as a continuation of Hiogo, and the two are active, thriving, and bustling; their narrow streets being thronged with people, kurumas, and ox-carts, while sweeping roofs of temples on heights and flats, torii, great bronze Buddhas, colossal stone lanterns give the native town a variety and picturesqueness very pleasing to the eye.
                                          Isabella Bird, *Unbeaten Tracks in Japan*, 216–17

This comes from Kobe ... the European portion of which is a raw American town. We have walked down the wide, naked streets between houses of "sham-dam" stucco, with Corinthian pillars of wood, wooden verandahs and piazzas, all stony grey beneath stony grey skies, and keeping guard over raw green saplings miscalled shade trees.

I faint in the Professor's arms. He had travelled a good deal. "This place is all right," said he. "It is Portland, Maine, but it is a bit too far west. Remarkably like Portland, Maine. You must go there." In truth Kobe is hideously American in externals. Even I, who have only seen pictures of America, recognise it at once. Like Nagasaki it lives among hills, but the hills are all scalped, and the general impression is of out-of-the-wayness.
                                  Rudyard Kipling, *Kipling's Japan – Collected Writings*, 52

Consider these two descriptions of settlement and life in Kōbe, Japan, by two of the more celebrated globetrotting British travellers of the 1880s. Kipling was twenty-four years old in 1889. Although born in Bombay, he had been raised in England. He had become one of the leading writers of his generation, noted particularly for the extent to which his writing explored and exposed complex issues of identity and national allegiance. Isabella Bird travelled to Japan in 1878. Already an adventurous tourist in her mid-forties, she had previously visited Australia, the United States, Hawaii, and Canada; on this excursion she would also visit China, Malaya, Singapore, and Vietnam; in later years she would visit India, Tibet, Morocco, Persia, Kurdistan, and Turkey.

Both writers focused on the Euro-American character of the settlement they encountered at Kōbe. Here in this place were familiar patterns: an ordered street layout and urban amenities, as well as the most modern apparatus of trade and commerce, including a railway linked to a steamship harbour. Both writers noted, too, a mixed population of Westerners, some of whom lived in suburban settings on higher ground above the town, in fashionable bungalows that drew from architectural precedents in England and New England. Here then is what can only be understood as a place embodiment of the global economy as it was emerging in the last decades of the nineteenth century. For Kipling and Bird, the landscape they confronted in Kōbe seemed a trifle sterile and lifeless. As members of a class apart from the world of trade, they seemed disappointed that this place seemed remote from the exotic world they had expected to encounter in Japan. Indeed, Kipling disparaged the American-ness of Kōbe and took pains to capture its slapdash character and to mock its self-conscious importance.

While the images of place contributed by Kipling and Bird are but snapshots, and were coloured by particularly narrow class biases, it is clear that the Kōbe they visited was yet another node in a broad and complex world economic system. Japan had resisted the full impact of colonialism by remaining its own political master; however, Kōbe and a number of other Japanese ports were part of an experience shared by a broad swath of the Pacific trading world. The presence in Kōbe of Euro-American traders, technocrats, and missionaries must be seen against a pattern of cultural transformation that would do much to reshape Japan. Moreover, what happened there was also occurring, or had already occurred, in Calcutta, Singapore, Hong Kong, Shanghai, and a host of other ports across the Asian Pacific space. This process had been under way for about two hundred years prior to the opening of Japan in the 1850s and had been noted as early as 1813 by William Milburn in his book *Oriental Commerce*.[1]

There is reason to see Kōbe's case as part of what Immanuel Wallerstein referred to as the emerging world economic system.[2] Critiques of Wallerstein's model have since developed a less structured one that allows more room for agency – a model perhaps exemplified by the Global Lives project of historical geographer Miles Ogborn.[3] This book follows Ogborn's approach in that it seeks to expose the human face of the Foreign Settlement at Kōbe as part of the "place making" and economic life of that port.

In the two centuries before 1850 the Japanese had deliberately cut themselves off from the world beyond their shores. The Tokugawa period began in 1603, and by 1624 the new rulers had expelled the

Spanish from Japan. This was followed by the Closed Country Edict of 1635, which prohibited Japanese from travelling outside the country on pain of expulsion or execution if they attempted to return. In 1638 the Portuguese left Japan, leaving only the Dutch traders at their tiny base at Dejima.[4] This policy of self-seclusion was steadfastly maintained until the mid-1850s. In opening itself again, Japan was re-entering a dynamic and complex world in which Euro-American political and mercantile forces exercised dominion over large parts of East Asia. Japan's transition into this system profoundly challenged its most basic cultural assumptions and indeed its autonomy, even as it faced the throes of an imperial regime change. The ascension of the young Emperor Meiji to the Chrysanthemum Throne in 1868 ushered in what would become for Japan an epic process of modernization and accommodation to outside forces. In the process, a handful of Japanese port cities would come to serve as the gateways to these external forces, which effectively made those ports laboratories for accommodating the Westerners and other Asians who served as agents of those changes. Kōbe was one such port, and by the end of the 1870s it was Japan's second port after Yokohama, as well as a stage where foreigners acted – often unwittingly – as exemplars of the modernity that so fascinated the Japanese. As we will see, Kōbe opened later than the other treaty ports grudgingly tolerated by the Japanese. It did so on the heels of a mercantile gold rush that for a moment in time in the 1860s generated great profits for those first on the scene at Yokohama and Nagasaki. By the time Kōbe opened itself to trade, the lustre of these economic boom times had diminished greatly, so that the 1870s would be what one anonymous observer called a "doubtful desert" – though as we will see, many Western traders continued to believe that a bonanza might yet be found in the "oasis" they were founding at Kōbe.[5]

This book sets out to reconstruct and illuminate that founding and in so doing to assess the ambiguities of the cultural encounter that was at the heart of the port's life. It is important to say a word about the limits of this study. To the extent that it emphasizes place making by Euro-Americans abroad, my research has focused on the documentary record that has survived in European languages. This is, however, a one-sided view, one that can be expanded only by those who are able to penetrate the Japanese-language documentation. My hope is that this book will engage both scholars of Asian economic and social history and geography, and readers with little or no prior knowledge of Japanese treaty port development. Throughout this book, I will be moving back and forth between micro and macro lenses – that is, between the broader narrative of regional and national development and an exploration of the lives and actions of the individuals who formed the foreign community of Kōbe. The aim in this is to illuminate a time and space; to make sense of the role played by Westerners in the larger transformation of Japan in the first three decades of the Meiji era, and to broaden our appreciation of the breadth of the Euro-American field of socio-economic operations in the Victorian era. For those of us descended, literally and figuratively, from the Anglo-European populations who settled overseas, it is worth being reminded that not everyone bent on remaking themselves in the "New World" found their opportunity on the resource frontier of the United States, or in the various other settler colonies of the Americas, southern Africa, or the Antipodes. Many participated instead in a mercantile world, principally as sojourners, but also as settlers, and in so doing they fabricated their own settlement template and landscape in the midst of urban societies that were the antithesis of the thinly populated frontiers that formed the base of so much of European settlement overseas. Much of this mercantile settlement was transient, in terms of the propensity of the participants to return home or to move from place to place in the far-flung mercantile network. There was also a tendency for the host society to blur the impact of these settlements by the sheer weight of its own population and subsequent post-colonial history. Not surprisingly, these settlements had largely faded from the Western imagination as the twentieth century unfolded. At best, our popular impressions of this episode in history survive in the form of literary caricatures, such as the romantic portrayals of the meeting of cultures in Puccini's opera, *Madam Butterfly*,[6] and in "blockbuster" historical novels such as James Clavell's *Gai-jin*, set in the treaty port of Yokohama ca. 1862, or his earlier *Taipan*, set in a related expatriate settlement, Hong Kong, in the 1840s. In reality, the men and women who founded these settlements lived comparatively ordinary lives, albeit in (for them) exotic settings. As we will see, many men did take Japanese "wives," and some pursued profits and power in the cut-throat world of commerce. But if caricatures there be, these treaty port residents were more aptly seen as ordinary individuals who saw in Japan an opportunity to pursue social and recreational goals that probably would have been above their station had they attempted to achieve them within the more rigid social boundaries of their home country. Thus, much of life in Kōbe was a relentless commercial routine, punctuated by the self-conscious affectations of upper-middle-class lifestyle aspirations and fantasies. The details of this life and the people who lived it will appear later.

The social and economic dynamics of the treaty port episode at Kōbe, as at the sister treaty ports of Yokohama and Nagasaki, created a strongly Euro-American–inspired human landscape to suit the foreign inhabitants of these places. The treaties restricted the foreign residents to these spaces, and as a result, they were surrounded by the rapidly urbanizing Japanese. The townscapes and land uses within the Foreign Concessions provided a measure of familiarity and comfort for the foreign occupants. To date there has been scant coverage of the process of place making that is the main subject of this study.

Acknowledgments

Research for this book has been carried out in Japan, Great Britain, the United States, Canada, and Australia. While some sources written in Japanese have been accessed with the help of translators (Ms Miyako Oe in Canada and Ms Noriko Kadota in Japan), most of the research has relied on documents written by English-speaking officials resident in Kōbe and published in newspapers there. I must also acknowledge a debt to Professor Hiroko Fujioka of Ritsumeikan University, Kyōto, for providing sound advice as well as documents arising from her research on Kōbe. At various times, in both Canada and Great Britain, I have been ably assisted in the transcription of documents by my daughter, Sarah Ennals, and it has been wonderful to share an enterprise and time with her. I owe a great debt to Kwansei Gakuin University in Nishinomiya, which twice has appointed me a Visiting Professor. It was through this contact with Japan that this project came into being, and it is to Professor George Oshima and KGU that the result is dedicated. Gratitude is also expressed here to the Social Sciences and Humanities Research Council of Canada for funding that enabled much of the archival work conducted abroad. Additional funding has been provided from the Marjorie Young Bell Faculty Fund of Mount Allison University.

    I am greatly indebted to a large number of colleagues who in various ways have influenced and encouraged me in this project. Dr Cecil Houston, former Dean of Arts and Social Science at the University of Windsor, planted the seed of interest in going to Japan in the first instance. When I arrived at Kwansei Gakuin University, Professor George Oshima, Professor of Geography, who then served as Dean of Humanities, provided a warm and tolerant setting where I could work and learn, as did many others at KGU on the occasions when I was privileged to be a visitor there, including Chancellor Mitsuo Miyata, and Chancellor Ichiro Yamauchi, President Ken Takeda, and President Hiroshi Imada, Professors Tsuneoshi Ukita, Yasuyuki Yagi, Shinpei Seguwa, the late Masahida Uchida, and the staff in the Office of International Programs, including Nobuo Tanii and the late Makoto Fujita. Thanks are also due to those with whom my family and I shared the experience of being *gaijin*, several of whom were missionary teachers working in a Japanese institution: Lynn and Masako Ryan, Mike and Kamiko Elliot, Donna and Shig Tatsuki, Gil and Maxine Bascome, Jim and Jean Joyce, Ruth and Mark Reames, Marci and Akiie Ninomiya, David and Sheila Woodsworth, and Ron and Camilla Jenkins. Closer to home, over many years, I have valued the friendship and intellectual stimulation of Graeme Wynn, Brian Osborne, Martyn Bowden, Tom McIlwraith, Ted Muller, Peter Rees, David Meyer, Joe Wood, Stephen Hornsby, Victor Konrad, Larry McCann, Stephen Bell, Gordon Winder, Rob Summerby-Murray, Eric Ross, Cole Harris, Jim Lemon, Ronald Rees, John Stanton, and Ian Newbould. Deryck Holdsworth, with whom I have enjoyed a wonderfully satisfying "writing partnership" since 1978, provided a thorough and challenging critique between the initial draft and the final product, for which I am deeply grateful. I also acknowledge the pithy critiques and suggestions of the external readers who reviewed the manuscript prior to publication. I am especially grateful also to Douglas Hildebrand, at the University of Toronto Press, who shepherded this project through to publication. All of these people have in their own way helped me better understand the nature of sound scholarship, cultural sensitivity, and academic integrity, and many have played no small role in pushing my specific research interests and capabilities toward a better understanding of urban places, wherever they may be. I owe a debt as well to the many unnamed archivists and librarians who eased my way at the Kōbe Municipal Archives, the National Library of Australia, the University of London, the Public Record Office, Kew, the Library of Congress, the Cambridge University Library, the Baker Library, Harvard University, the Peabody Museum, and the University of Toronto. I acknowledge with gratitude the Geography Department at the Australian National University, which enabled me to be a short-term Visiting Fellow and to join the community of University House during my time in Canberra. This project has moved in fits and starts, many of the fits the result of having served first as Dean of Social Science and later as Vice President, Academic and Research, at Mount Allison University. University administration demands much of one's time and emotional energies, and finding the opportunity to maintain momentum on a project such as this seemed all but impossible. However, the rewards of service have been two administrative leaves, first in 1997–8 and again in 2002–3, and the blessed freedom of these periods allowed this project to reach fruition. For this I am grateful to Mount Allison University. Finally, the opportunity to share a year in Japan with my family – Cheryl, Sarah, and Andrew – was a rare privilege that enriched our lives in many and enduring ways. But it is Cheryl who has borne the travails of this long project with unstinting grace and confidence in me, and it is to her that I lovingly reserve my greatest thanks.

OPENING A WINDOW TO THE WEST

The Foreign Concession at Kōbe, Japan, 1868–1899

Chapter One

Setting the Stage: The Role of Ports in the Encounter between East and West in Japan

Along the coasts of South and East Asia are a number of ports with long-established connections to east Africa, the Persian Gulf, and the eastern Mediterranean.[1] For example, Mombasa in what is now Kenya has a trading history going back to the twelfth century, and Kochi (also spelled Chochi) in what is now the Indian state of Kerala attracted Jewish resident traders as early as 352 BC, as well as Muslim merchants in the eighth century, and was a regular port of call in the spice trade from the fourteenth century onwards. The Chinese had penetrated the Indian Ocean by 1403, when a fleet of "treasure ships" under Admiral Zeng set out to assert the commercial and imperial interests of the Ming dynasty. In the sixteenth century, this trading zone drew the interest of Armenian and Portuguese traders, who succeeded before other Europeans in penetrating the Indian Ocean realm, as evidenced by the Portuguese conquest of Goa on the west coast of India in 1511.[2] The Portuguese also penetrated the East Indies, establishing a base at Macao on the south China coast in 1510 and conquering Malacca on the Malay Peninsula in 1511. From these places, they conducted their lucrative trading activities and set out to convert the Asian "heathens." Portuguese missionaries, members of what would become the Society of Jesus (the Jesuits), made a concerted effort to win converts throughout southeastern Asia, and some of them reached Japan in 1543. St Francis Xavier, a co-founder of the Jesuit order, was himself in Japan in 1549–50, and for a time, Christianity made strong inroads in western Japan. These missionary activities led to an important period of trading contact known in Japan as the *Nanban* ("southern barbarian") trading period, during which the Japanese engaged the Portuguese as well as, after 1609, the Dutch through the Dutch East India Company

(*Vereenigde Oost Indische Companie*; VOC). This relationship was predicated on an exchange of goods, particularly textiles such as cotton and silk, but also on cultural ideas and on technologies, particularly with respect to shipbuilding, military crafts, and the arts. The British East India Company briefly operated a factory (i.e., trading post) at Hirado in what is now Nagasaki prefecture between 1613 and 1623 under Richard Cocks; these efforts involved the "larger than life" activities of William Adams, who became close to the Shōgun and helped transmit British navigational and shipbuilding knowledge to the Japanese.[3]

These trade initiatives did not last long, however; in 1638, after the founding of the Tokugawa civil dynasty, the foreigners were ejected as part of a wider project of national unification.[4] Under the tight military regime of the Tokugawa Shōgunate, Europeans were proscribed thereafter from having contact with the Japanese. While it is generally assumed that some incidental trade and contact took place around the edges of Japan, the only Europeans with an officially sanctioned window into Japan's national space were the Dutch, through their outpost on the artificial island of Dejima (also spelled Deshima) in Nagasaki harbour. Here the Japanese systematically and selectively extracted Western scientific and other knowledge in exchange for certain goods such as ceramics.

In the early nineteenth century the Dutch, English, and Portuguese, now joined by the French, Russians, Germans, and Americans, coveted the wealth to be won through trade in East Asian waters. During the 1840s, despite the massive collective and not insignificant military resources of the Qing dynasty (1644–1912), China was pried open by a combination of British naval power and commercial chicanery; the British did not hesitate to traffic opium, which they used to pay for the tea that China exported to Britain. The British had already gained limited trading access to southern China through the port of Canton, where the East India Company had been permitted to establish a factory. But the limits placed on their freedom of movement on Chinese soil, and disputes over which nation's laws would prevail in cases involving British nationals in China, clearly frustrated the company and its political supporters in Britain.[5] Britain responded in 1839 by seizing Hong Kong, which hastened the erosion of relations between Britain and China so that war became inevitable.

Following their victory in the Opium War of 1839–42, the British exacted a second humiliating concession from China: five more ports were opened to the West in 1843, the most important of which was

Shanghai. That port was already a trading hub, with a large fleet of junks linking it to other ports both north and south. But more than this, Shanghai was a gateway into the heart of China, as far as Wanxian in Sichuan province, along the Yangtze River. For these reasons Shanghai quickly superseded Canton as the centre of Sino-Western trade.[6] Shanghai was significant for another reason: under the terms of the political concessions extracted by the British in 1842, it was to be opened to other Europeans under a treaty relationship that granted Westerners territory within the city. Within this Foreign Concession, foreign nationals, under their respective consuls, would be subject to European rather than Chinese laws. Westerners viewed this arrangement as essential, for they abhorred the intractable judicial processes of the Chinese and especially their propensity to apply torture as part of punishment.

Note that the British had by then established their own precedents for this type of ethnocentric legal imposition: in India, where Mayor's Courts had been established in Bombay, Calcutta, and Madras that upheld English common law, with pre-emptive authority in the hands of the East India Company beginning in the 1720s, and then in those of British Imperial authorities after the 1770s.[7] Not surprisingly, British traders and their European rivals assumed that they would enjoy similar legal rights elsewhere in Asia. There was also a practical economic rationale for this arrangement, which evolved out of the longer history of British and European trade with China. Under the Chinese system, imported goods were subject to an entry tariff; then, as they were transported into the hinterland, they were subject to further charges and taxes. In this way, the country in matters of trade was more like a customs union than a centrally controlled entity. By the terms of the treaty port agreement, goods entering China en route to destinations beyond the port of entry were exempted from any further charges. Trade through non-treaty ports, however, would continue to face the problem of further up-country charges.[8]

This "extraterritorial" relationship would serve as an important precedent for the foreign powers when they penetrated Japan. The British, through the East India Company, had by 1850 developed a model for conducting trade that served their national interests. Critical to that model's development was the British experience "on the ground" in India, where the "presidency cities" of Madras and Calcutta had been segregated into "white and black towns," thereby setting a pattern for racially distinct enclaves.[9] In Shanghai the British arranged for a Foreign Concession that in the end would separate their nationals not only from the Chinese but also from rival traders who were foreign nationals – namely, the French and the Americans.

Japan in 1850 was the final frontier for Euro-American commerce and culture in Asia. The Shōguns had up until then resisted all efforts by Westerners to penetrate their secluded island realm. In actual fact, the Japanese relationship with the outside world had been a cat-and-mouse game for centuries: there is now a good deal of evidence that Japan had long engaged in trading relationships with its Asian neighbours, despite its embargo on such activity. Indeed, before it enacted its isolationist policy, Japan had for a time seemed poised to become a colonial and commercial force in its own right, in Formosa and the Philippines.[10] However, the Japanese seem to have developed a strongly geocentric habit of mind. And, as Yonemoto points out, they were also astonishingly poor navigators and cartographers, which weakened their strategic position in the Asia-Pacific orbit.[11]

Japan's self-seclusion under the Tokugawa Shōgunate owed much to its particular domestic problems that were consequences of a sophisticated yet fractious feudal order. But it was also bound up with a profound suspicion of outsiders. The presence of Portuguese missionaries in Japan seems to have been an especially distressing episode, one that resulted in the state exercising its considerable powers to brutally stamp out Christianity. The Tokugawa Shōgunate, very soon after wresting control of the country in the 1630s, in an effort to secure domestic stability, decreed Japan's isolation. In addition to preventing the entry of foreigners onto Japanese soil, that policy restricted the movement of Japanese nationals abroad. The paranoia felt by the Tokugawa Shōguns was rooted in their fear that their political rivals would procure superior military armaments from conniving Europeans. Nevertheless, one window onto the outside world was tolerated: in 1641, the Dutch East India Company was allowed to establish a small trading base on the island of Dejima in Nagasaki Bay. There, under strict controls that forbade direct contact between ordinary Japanese and the Dutch, authorities permitted what for two centuries would be the only real mercantile and intellectual contact between Japan and Europe.

Credit for finally breaking Japan's claustrophobic isolation is conventionally given to Commodore Matthew Perry of the U.S. Navy, who arrived in Japan in the early 1850s with a small expeditionary fleet. But scholars have documented a long series of encounters and episodes that brought pressure to bear on the isolationist policy of the Bakufu[12] in the decades leading up to the arrival of Perry's "Black Ships." The remarkable British trader and administrator, Sir Thomas Stamford Raffles, while Governor of Java, had attempted to penetrate Japan and break the Dutch monopoly in 1813 and 1814. From time to time, British and other survey ships had put into Japanese harbours to escape bad weather or to take on fresh water – occasions that often earned them a less than courteous welcome. Nor was humanitarianism always evident, for example, in the repeated cases of sailors – many of them from whaling expeditions in the North Pacific – who found themselves shipwrecked along the Japanese coast. The directives given to Japanese officials were such that castaways were often harassed and imprisoned, although some local officials were more gentle. Eventually, however, a certain quid pro quo developed as Western traders sought to repatriate shipwrecked Japanese sailors from time to time. By the final decades of the eighteenth century, a more sustained foreign threat to Japan was emerging, in the form of expanding Russian settlements and fishing activities in the Kurile Islands on Japan's northern flank.[13]

Pressures to confront the outside world were building within Japan as well. In the second quarter of the nineteenth century, internal forces for reform were pressing against an increasingly tired and weak Bakufu. Several internal conditions hastened this. These included the cumulative impact of the monetary devaluations that the Bakufu had been declaring in order to remain solvent after years of profligate expenditures.[14] These devaluations placed a heavy burden on the common people, who felt themselves being crushed by the taxes they were having to pay. A serious famine in the 1830s – the Tempo famine – exposed the regime to further pressure as a result of its failure to provide effective relief. It is apparent, too, that the long period of seclusion was only whetting the curiosity of many of Japan's brightest, who hungrily devoured the small bits of foreign knowledge, called *rangaku*, that were filtering in through Japan's contacts with the Dutch. Finally, around this time the regional forces that had long been held in check by the Bakufu's central military regime were beginning to surface again. Especially in Japan's peripheral regions, forces were at work that would, when the opportune moment came, challenge the very existence of the Tokugawa regime.[15]

The arrival of American fleet in Tōkyō Bay in 1853, with its flagrant show of force, served as the catalyst for profound change. Fear of superior Western armaments and the resolve of Perry and his diplomatic successor, the remarkable Townsend Harris, caused an increasingly weak and uncertain Bakufu leadership to blink. On his second visit to Japan, Perry brought a larger and even more powerful squadron, which enabled him in March 1854 to persuade the Japanese authorities to conclude a treaty opening two ports, Shimoda and Hakodate, to American trade (see Map 1.1).[16] The treaty further specified that there would be American consular representation in Japan. In October 1854, Japan signed a similar treaty with the British. This was followed with bilateral treaty arrangements with Russia in February 1855 and with the Netherlands in November 1855.

There followed a period of internal confusion, brought on in part by a series of succession crises within the Shōgunate. During this same period the American Consul, Townsend Harris, through sheer personal fortitude, succeeded in gaining further concessions, including the opening of the port at Nagasaki in 1857. But Harris's more significant achievement was a comprehensive treaty, concluded in July 1858, that promised the Americans full diplomatic and consular privileges, the opening of more ports to U.S. foreign trade by agreed dates, the establishment of private import and export merchants in those ports, and the setting of agreed tariffs. This agreement also ensured freedom of religion for the Americans and detailed the extraterritorial jurisdictions and protocols for consuls and their agents. Over the next few months, the British, French, Dutch, and Russians would conclude similar treaties. The British signed what became known as the Elgin Treaty or the Anglo-Japanese Treaty of Amity and Commerce, which called for the opening of selected Japanese ports to British traders and their vessels. Hakodate, Kanagawa, and Nagasaki were scheduled to open on 1 July 1859, Niigata a year later, and Hiōgo (Kōbe) on 1 January 1863.

All of this was a remarkable about-face for Japan: at a stroke, the Japanese, long cut off and self-preoccupied, had been made to push back their screen of seclusion and reveal themselves to the world. At the same time, they began to absorb foreign ideas and techniques that to them must have seemed revolutionary. In the process, the treaty ports were thrust to centre stage as Japan's windows on the outside world. Historians are now generally agreed that Japan was ripe for change.[17] But this apparent new spirit of openness was, of course, in many respects illusory. The opening had been largely forced on them, even though that coercion was more implied than demonstrated. Remember that despite their long isolation, the Japanese had been strongly aware of the remarkable capabilities of Western armaments. Moreover, the opening had been a humiliation for the Bakufu, and the authorities continued to be highly wary of foreigners and to exert a strong measure of control over them. As we shall see in the case of Kōbe, while they had granted a few treaty ports, the Japanese would severely restrict the movements of newcomers within the country; and even within the treaty ports, foreigners would find themselves restricted to rather uninviting compounds set aside for them. Moreover, the Japanese authorities were still in sufficient control to exercise their well-cultivated mastery of evasion, obfuscation, and vacillation on important matters – an art that few Western diplomats or traders understood or seemed able to accommodate.

It is unclear how Western governments perceived the treaty ports in terms of advantages. In hindsight, the prize locations were Kanagawa and Hiōgo, which served Japan's two main population centres. Nagasaki and Hakodate would enjoy modest success as trading centres; Shimoda and Niigata would be dismal failures. Kanagawa was the port for Yedo (Tōkyō),[18] which was the political centre of the nation under the Shōgunate and which would maintain that function even after the collapse of the Shōgunate and its replacement by a new political order under the Emperor Meiji.[19] Hiōgo, on the Seto Sea (a.k.a. the Inland Sea), had long been an important port on the island of Honshū, for it was adjacent to the already important manufacturing centre of Ōsaka, whose own port, like that of Tōkyō, suffered from navigational difficulties. Hiōgo was also the nearest port to Kyōto, the seat of the emperor and centre of the nation's traditional culture. Like other major metropolitan seats of power in East Asia, Kyōto was located inland.[20]

Hiōgo's proximity to Kyōto was the rationale for a last desperate effort to frustrate the foreign intruders: the Japanese Shōgunate requested that the opening of Hiōgo be delayed for five years. This request had its roots in a struggle within the feudal order. Some within that order rallied against the foreigners, no doubt because it was the hated Shōgunate that had bowed to the Westerners' demands. These forces of opposition were themselves complex, however. The powerful Satsuma clan were long-time enemies of the Bakufu as well as the leading advocates for the emperor's restoration to a central position in governance. Satsuma, and other feudal clans similarly at odds with the Bakufu, worried that the opening of Hiōgo – and of Ōsaka, which had been added to the list of ports to be opened – would provide the Shōgunate with increased revenue from customs duties and thus save it from the financial collapse that threatened it.[21] In marshalling their case, the dissidents seized on the prospect of a foreign settlement located so close to Kyōto, seat of the emperor. Such foreign proximity, they argued, would threaten the sanctity of the imperial household. The foreign powers, realizing that the country was sliding into chaos, and having already experienced the murderous fury of anti-foreign elements, chose not to press the issue.[22] No doubt the building of the apparatus of trade at Yokohama and Nagasaki and the other open ports was enough of a challenge for the time being. They knew that the day would come when Hiōgo and Ōsaka were thrown open to foreign settlement.

Let us return to the matter of the Western penetration of East Asia, focusing on the cultural dynamic at work and its implications for the societies confronting one another in the ports of this part of the world. From Jeddah to Bombay, Madras, Colombo, and Rangoon, from Batavia, Singapore, and Bangkok, and thence on to the China coast, Europeans, particularly the Portuguese, Dutch, and British, had over three centuries expanded their respective orbits of trade and in so doing had exerted extraordinarily powerful forces that transformed these centres. Through commerce, missionary activity, and inevitably colonialism, they had infiltrated these ports, insinuating their agents in such a way that these centres became the beachheads for what would be a profound cultural assault on these societies. Cultural hybrids, these ports accommodated the complex apparatus of trade while serving as homes away from home for a host of foreigners. Not surprisingly, the result was a distinctive port landscape. At the same time, these centres became portals through which Western influences passed into the hinterlands and from which knowledge of Asian ways was carried back to the West.

Social scientists have been exploring the modern dynamics of these cities, but few historians and perhaps even fewer historical geographers have looked deep into this trading world of the past. In Western scholarly writing, note must be made of the prodigious scholarship on the treaty ports of China prepared by two Americans: historian John King Fairbank and geographer Rhoads Murphy, whose respective assessments of mid-nineteenth-century Shanghai are still remarkably fresh and insightful. But both represent the contribution of a generation of scholars working in the middle decades of the twentieth century.[23] More recent scholarly interest, captured in *Brides of the Sea*, edited by Frank Broeze of the University of Western Australia, suggests that these cities and the trading relationships that linked them merit renewed attention. In the late twentieth century, as the ghosts of the War in the Pacific and the Cold War were shaken off, East Asia regained political, economic, and intellectual prominence, and the West has had to come to terms imaginatively and practically with the sheer population and economic potential of this part of the world. Historical scholarship invariably follows rather than leads, so it is perhaps not surprising that these cities and their regions are now being "discovered" by a new generation of scholars.

The editors of *Brides of the Sea* lamented the lack of a comparable work on Japanese trading ports during the Meiji period, especially since they had been port cities before the European arrival and thus were comparable to other such cities in Asia.[24] Few Western scholars – and, it seems, few Japanese ones – have conducted detailed studies of Japan's urban history in general, let alone individual cities.[25] Since Broeze's volume, James Hoare's study of the Japanese treaty ports has been published.[26] That work, which began as a dissertation at the University of London in 1971, fell dormant after Hoare joined the diplomatic corps. Its publication in 1994 was a welcome addition to the literature in English on the Japanese treaty ports. By painting a broad picture of the forces that produced those ports and by exploring how foreign newspapers provided models for an indigenous Japanese press in the Meiji period, Hoare has established a solid base for a detailed exploration of particular treaty ports in Japan.

To this can now be added Kevin Murphy's fine work, *The American Merchant Experience in Nineteenth Century Japan*.[27] Murphy examines the origins and behaviour of American merchants – primarily those in Yokohama, although he also relies heavily on evidence that survives from Kōbe merchants. His work is especially good at situating American merchants in a particular social-behavioural context that reflected post–Civil War American male values. Another welcome and valuable addition to the literature is Yuki Allyson Honjo's *Japan's Early Experience of Contract Management in the Treaty Ports*.[28] That study focuses on the struggles of Western traders and Japanese merchants to adapt to the legal frameworks of international trade where none had existed. Honjo explores how the two sides sought to develop mutual trust, trade logistics, and modes of credit and how the Japanese wrestled with the creation of specialized courts for legal recourse, again primarily for Yokohama during the early years of the treaties, 1859 to 1871.

The present study considers in depth the Western penetration of one important Japanese port city during the Meiji period. Focusing on the Foreign Concession at Kōbe will allow an attempt to measure the impact of Japan's controlled accommodation to the West. It also enables us to examine the general pattern of Asian port development as well as cultural penetration and transformation, the more so because the Japanese treaty ports were the last to be so affected besides being at the outer geographic limit of the Euro-American trading system. Indeed, Kōbe was among the last Japanese treaty ports to be opened, although it quickly became second only to Yokohama in terms of economic importance and scale of foreign presence. There is no scholarly study in English of Kōbe during its years as a treaty port. However, because it remained an important base for foreign trade and foreign corporate concentration, it retained a vital and (one might even say) vibrant expatriate community throughout the twentieth century. One member of that community, Australian-born Harold S. Williams, has produced an impressive body of "local history" pertaining to the treaty port era, originally published in the form of regular features in the *Mainichi Shinbun*, the local English-language newspaper. Those features have been the basis of a number of books.[29]

By providing a scholarly treatment of Kōbe, this volume contributes to broader empirical considerations of Japan's economic leap forward toward modernity. Such discussions have preoccupied scholars in Japan and the West, who have sought to establish whether Japan's rapid emergence as an advanced industrial economy offers lessons for other developing countries. This discussion has revolved largely around Japan's predisposition for modernity in social and political but primarily in economic terms, as a consequence of developments during the long Tokugawa period. Many scholars have tended to draw a sharp line between the agrarian Tokugawa and the capitalistic and industrial Meiji eras, with the purposeful efforts of the Meiji leaders representing a sharp break from older feudal patterns. This interpretation implies a parallel with developments in Western societies, albeit sped up, given that Japan's leap was seemingly made within a couple of generations. According to this Western-centric line of argument, the insertion of Western agency in the form of the treaty ports and the pursuit of trade, the recruitment of Western experts, and the importation of Western technology, governance models, and legal systems, were all crucial to Japan's inevitable modernization.

More recent scholarship has challenged this line of thinking by seeing the Meiji period more as a political watershed and by emphasizing the continuity between the socio-economic changes of the Tokugawa period and the industrialization project of the Meiji era. Some scholars who have analysed the "seclusion" or "isolation" policy of the Tokugawa

Bakufu suggest that there were strong efforts to establish Tokugawa legitimacy not just within Japan but also in other parts of East Asia.[30] Others have drawn attention to the scale and importance of foreign trade during the "seclusion" period, noting that while the Japanese authorities severely limited trade, still there *were* exports, especially of silver, which continued in large quantities up until the mid-eighteenth century, conducted through Korea and the Ryūkū Islands, and that Japan formed part of an East Asian trading bloc centred on China.[31] There was no direct link between this trade and that ushered in during the mid-nineteenth century; even so, our understanding of Japan's "seclusion" policy has necessarily undergone considerable modification as a result of these findings.

The Nature and Structure of Ports

In the first half of the twentieth century, academic geographers had an interest in deconstructing the physical, social, and economic characteristics of ports, with particular attention paid to the morphology of the city itself in distinguishing ports as particular urban types.[32] This type of functional and morphologically based analysis gave way in the second half of the century to a fascination with more sociologically rooted types of research, which grappled mainly with a range of twentieth-century dynamics, such as the rise of the automobile and new modes of production that reshaped cities into metropolitan regions marked by unprecedented levels of social sorting based on class and race, especially in North America and Europe. Not surprisingly, those who focused on urban history during this period reflected these lines of inquiry. More recently a handful of scholars have been revisiting these earlier approaches, with the result that new insights are emerging from the blending of social, ethnographical, and functional analyses. The present study aims to build on to this literature, using Kōbe as an example of the hybridization that resulted when Western economic forces arrived on the ground in urban spaces that had previously been patterned by other cultures' political, economic, and social determinants.

Ports are essential portals between a given region's economy and the world beyond. It follows that the development of a successful mercantile trading economy is conditioned by a number of factors, both human and physical. Attributes such of the scale and navigational features of the harbour are one factor (in this regard, many ports have relatively poor harbours). As the size and draught of oceangoing ships increased, only those ports with harbours capable of accommodating those vessels were able to retain and further exploit their functions. Even good harbours had to be dredged, and increasingly ports needed to invest significantly in new and more efficient wharfs, loading facilities, and navigation aids. The ideal port offers easy entry for vessels, ample anchorage, deep water, and a small tidal range (or constancy of water level if on a river); it also offers shelter from damaging storms and a climate favouring year-round operations (e.g., it is free of ice all year). Deficiencies regarding these ideals can often be overcome, or mitigated by remedial engineering such as breakwaters, dredging, and tidal gates.

To these might be added the port city's site and location and whether these favour both urban settlement and transport connections to a productive hinterland. Just as the physical attributes of the harbour are crucial to a port's success, so too is the quality of the landward transportation system along which trade goods flow to and from the port. Before the railway age, most trade goods were transported overland by road or along rivers flowing into the port. Also, goods destined for export might move along the coast on local vessels, which converged on the dominant port. The converse applied to imported goods, which were distributed from the port by coastal vessels. Such coastal activity extended the catchment area for a port's trade, an area often referred to as the port's *foreland*. Complicating this picture, two or more ports might share a hinterland and foreland and thus compete for trade. This could lead to specialization of trade based on a comparative advantage related to certain goods or services. The advantage could arise from a particular specialized clustering of merchants, financial guilds, kin networks, or domestic and/or overseas connections.

Ports typically derive their raison d'être from their "break bulk" factor – in other words, from their capacity to move goods from one form of transport to another at the point where land and water transport meet. By minimizing loading and unloading costs, as in the case of bulk cargoes (e.g., grain, ore, crude oil), or by adding value through the refining of raw materials, or better still, by turning those materials into finished goods, ports create cost efficiencies that are vital to their economic existence. The latter activities have encouraged port cities to become processing and manufacturing centres.

Morphologically, ports begin at the edge of the harbour where land and water meet. Over time, docks, warehouses, chandlers, and services for sailors come to be clustered at this location, where they form a distinctive landscape associated with ports. Housing and basic services for the local residents may also be clustered around the harbour. Typically, these include the homes and offices of merchants and shippers; as trade increases, these may be joined by banks, insurance offices, assayers, customs officials, police, and so on.

Many ports are cosmopolitan. Being open to the world inevitably leads to a varied mixture of races, cultures, and ideas, and this results in cultural hybridization. This factor has added a sociological and physical dynamism to the life of ports that finds expression in everything from the blending of races to the creation of unique architectural forms.

Let us consider Kōbe with respect to these morphological patterns and issues. First of all, it is difficult from the record to reconstruct Kōbe's hinterland with great precision since records of the trade goods procured by foreign merchants through Japanese merchants and intermediaries are not part of the record examined in this study. Thus what follows is a conjectural reconstruction of what the hinterland probably was, based on geography as well as on agricultural and production patterns as they relate to key goods.

As noted, the port of Kōbe was, paradoxically, both distinguishable and indistinguishable from its immediate neighbour, the port of Hiōgo. At the time that Kōbe was declared open in 1868, Hiōgo was a city of some 10,000 people and had served as a port for both internal and external trade for several centuries. Kōbe, which was the Foreign Concession of Hiōgo, would never reach a population greater than 5,000 during the treaty port era. It was in reality an appendage of the port of Hiōgo, albeit one that, by virtue of its legal definition, its mixture of foreign occupants and their businesses, its separate customs house, and its singular focus on the support and execution of international trade, would become a specialized mercantile centre for a broader region. In this latter sense it would come to be seen as a port unto itself, both functionally and figuratively.

Kōbe gained its prominence and capacity to fulfil this role from the comparative advantage of its large deepwater harbour, as well as from the entrepreneurial, trade, and financial expertise of those who gravitated to it from the moment it opened. Here it must be noted that Ōsaka, a city of some 100,000 people in 1868 and the most important industrial centre in all Japan, lay only 20 kilometres to the east of Kōbe. But Ōsaka's lack of a harbour capable of handling oceangoing ships, and the absence of expertise among its merchants that would have enabled them to immediately exploit the opportunities of international trade, meant that Kōbe's merchants enjoyed a comparative advantage that overrode Ōsaka's overwhelming scale and traditional economic power. Simply put, Kōbe was circumstantially an anachronism with respect to its scale and location, but its purposeful role as a treaty port, and the comparative advantages it enjoyed as result, made it distinguishable as a port of substance and power in this period. This is all the more remarkable given that it seemingly lacked some of the advantages of other major ports in East Asia and elsewhere.

Unlike Canton, Shanghai, and many other major ports, Kōbe was not situated at the mouth of a major river flowing from a rich and expansive hinterland. The rivers flanking Kōbe were short and essentially unnavigable. This meant that commodities and trade goods moving into and from the hinterland did so by other means, and it is clear that the coastal foreland had a strong impact on Kōbe as it extended its source region for domestic supplies and trade commodities. Kōbe, that is, could source its goods from around the Inland Sea, including the cities of Ōkayama and Hiroshima, from the island of Shikoku to the west, and from Ise Bay to the east – perhaps from as far as Nagoya (see Map 0.2). These far-reaching foreland connections enabled Kōbe to exploit the trading opportunities offered by goods such as rice, vegetable wax, seaweed, and other marine by-products. These goods were traded within Japan as well as along the China coast, particularly in Shanghai, Foochow, Ningpo, Amoy, and Hong Kong – the five treaty ports of China.

Kōbe's natural landward hinterland extended through Ōsaka. By means of one of that city's major navigable river systems, the Yodo and its tributaries, the Katsura, Kamo, and Hozu, it also tapped the territory surrounding Kyōto. Indeed, from the Kamo River, it was only around 10 kilometres overland to the city of Ōtsu on Lake Biwa, and this circumstance extended the navigable hinterland some distance northward into what would become Shiga prefecture, almost to the Sea of Japan coast. By means of the Yamato River, which also ran through Ōsaka, Kōbe's hinterland tapped the rich interior lowland surrounding the city of Nara and much of Mie prefecture (see Map 0.2).

Perhaps the most critical boundary for Kōbe was the one it shared with its chief trading rival, the treaty port of Yokohama. Especially critical in this competition was the flow of tea and silk, the two commodities that would dominate trade at both ports. At the time, both tea and silk production favoured Yokohama, since tea was a prominent feature of the area around Shizuoka, immediately to the west of Yokohama, and silk production was strongly centred in the Nagano region, northeast of Yokohama. However, there was some sericulture near Kyōto around the city of Ōtsu, and tea was grown in the uplands around the city of

Nara, thus providing Kōbe with access to producing regions within its own hinterland. These areas probably did not rival those of Yokohama in scale and quality of production, but they did ensure that Kōbe's traders had access to the two key staples of Japan's early export trade.

Overall, then, Kōbe was adequately placed for success by virtue of its hinterland and the quality of its harbour. The presence of a critical mass of foreign traders, who enjoyed a range of services provided by other foreigners as well as by the indigenous populations of Hiōgo and Ōsaka, ensured that Kōbe would become the second leading port in Meiji Japan. The development of tea processing in Kōbe, and the manufacturing of other key products – namely, ships and paper – suggest that at Kōbe, the break-bulk factor was at work and that key staples such as tea had "value added," as is typical of successful ports.

We now explore more closely the morphology of the pre-industrial mercantile city. That urban form emerged as a distinctive type in northwestern Europe during the sixteenth century, coinciding with the European overseas expansion.[33] A sizeable body of literature now exists to suggest that European expansion to North America resulted in what are conventionally termed "mercantile cities," with their colonial impetus implied. Meanwhile, another group of overseas mercantile cities, those in Africa and Asia, have generally been referred to as "colonial port cities." The distinction hinges on the fact that in these latter cities, a small European population operated a "firm-centred economy," while a large native population conducted a bazaar economy; this invariably established a racial, cultural and economic divide on the ground. In the twentieth century, the resulting cultural bifurcation was still discernible in the port cities of Southeast Asia. As a consequence, land use patterns took the shape of concentric arcs, which were broken by wedges that distinguished "alien" from "native" commercial and residential zones.[34] A second spatial model, this one proposed by Kosambi and Brush and based on the "presidency towns" of Madras, Calcutta, and Bombay, offers a more detailed portrayal of the dynamics and stages of growth of these colonial port cities.[35] Three stages of growth are noted: fortified trading posts were established; this was followed by movement inland as the colonial authorities established military, transport, governmental, and residential enclaves; and, finally, these enclaves gradually became more clearly defined and more exclusively European. Hornsby provides a glimpse of this phenomenon in his exploration of Calcutta, using maps of economic land use as one means to chart this development (see Map 1.1).

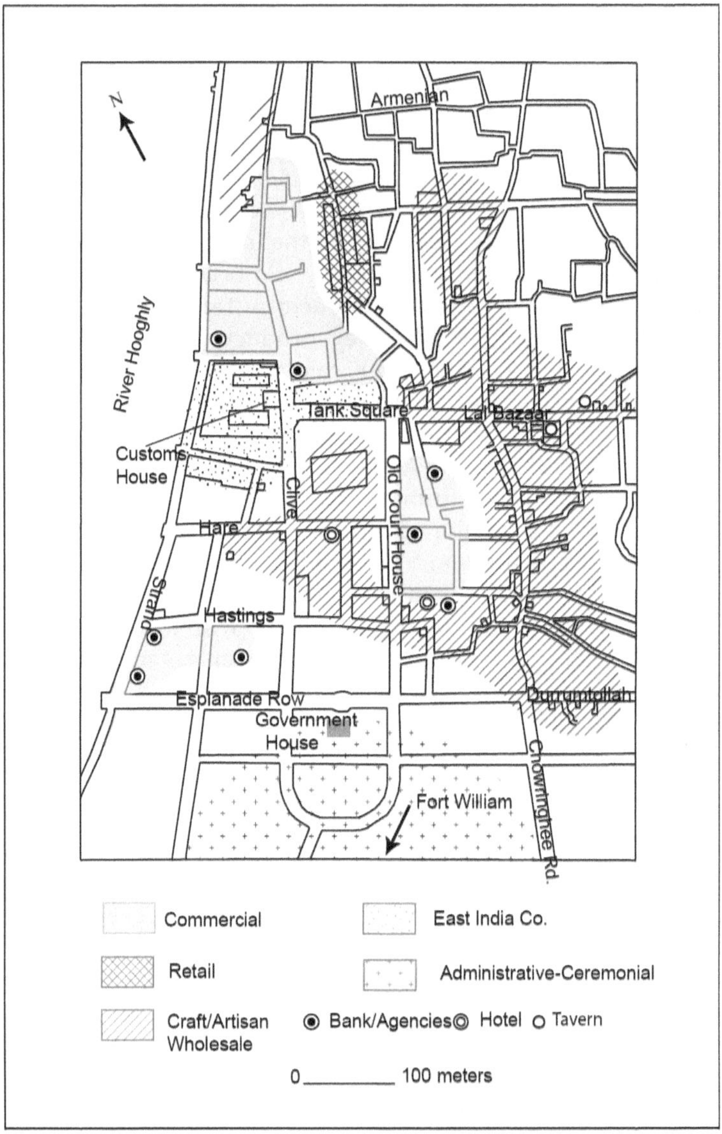

18   Opening a Window to the West

Map 1.1. Economic land use in early nineteenth-century Calcutta.
Source: Stephen J. Hornsby, "Discovering the Mercantile City in South Asia: The Example of Early-Nineteenth-Century Calcutta," *Journal of Historical Geography* 23, no. 2 (1997): 142. Reproduced with permission.

Similarly conceived spatial models of the North American mercantile city offer further insight into the process of city making in a different colonial context.[36] Martyn Bowden has proposed a triangular model for the North American mercantile city that employs a version of a capitalist land rent process such that activities were spatially segregated based on competition and complementarity.[37] The rent gradient meant that high land rent activities were in the city centre and low rent activities on the periphery. Accordingly, given the mercantile nature of these cities, the waterfront, including access to it, was of highest value. But here too there was a gradient of sorts, with the centre or the best sites on the waterfront being commanded by importers and exporters of high-value goods, and by wharfs for low-value goods, with related maritime activities such as shipbuilding relegated to the waterfront's periphery, where land was cheaper. The waterfront district thus exhibited a range of development activities, with the "centre" mixing the offices of commission agents, and shipping offices and their warehouses; farther out was a district dominated by merchant exchanges, banks, and insurance offices. Farther inland were two zones providing services that supported the resident and transient mercantile populations. Here were found retail establishments, hotels, social clubs, and recreational activities. At the apex of Bowden's triangle was a zone where political and religious affairs were situated, beyond which upper-class residential areas might be located. Establishments that were unable to pay high rents tended to be cast outside the triangle, on the flanks. These might include manufacturers, artisans' workshops, lower-class hostels serving sailors, and working-class neighbourhoods. Figure 1.1 offers a graphic depiction of Bowden's model.

To this picture might be added the three edifices that symbolically marked the points of the triangle: the three "Cs," customs house, courthouse, and church (or cathedral). Spatially, these points were situated so that the church stood as the landward apex, with the customs house and courthouse along the water. Each point carried clear cultural symbolism: the church denoting cultural continuity and moral authority, particularly in colonial settings, where its visibility and formality of architecture also symbolized colonial civilization; the courthouse denoting the rule of law and civil authority, which might also use architecture as a statement of a civilizing order; and the customs house denoting the centrality of trade and the regulated conduct of mercantilism as the economic engine of the city.

Figure 1.1. Bowden's Mercantile Triangle model.

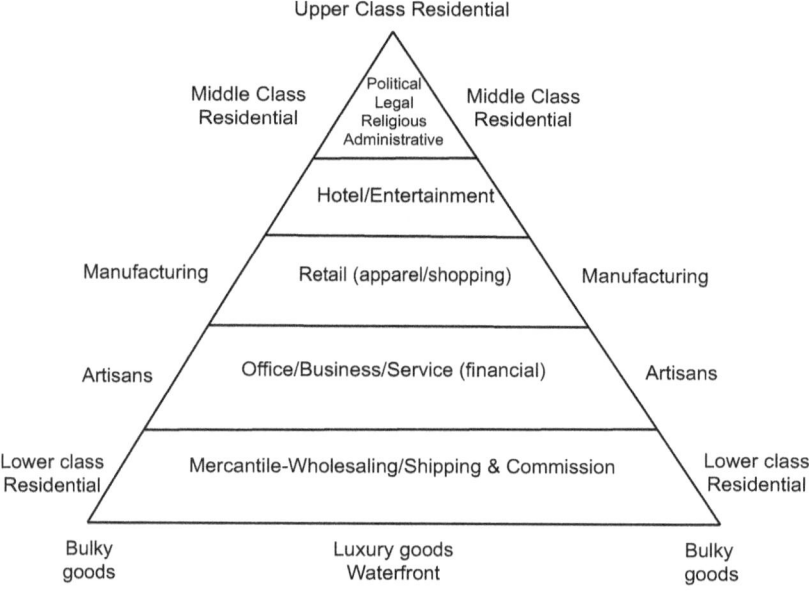

Source: Marin Bowden, Eastern Historical Geography Association Meeting Field Trip Guide (1989). Used with permission of the author.

What can probably be called the final phase of "colonial mercantilism" was carried out in the Far East for at least three or four decades after 1860 with the building of new ports and beachheads for overseas commerce. Kōbe between 1868 and 1899 was clearly such a beachhead, as were Shanghai, Hong Kong, and Yokohama. These places formed what might be regarded as the "last frontier" in the longer episode of mercantile port making. Did the treaty port at Kōbe reflect the triangular pattern of land use described above? Kōbe, like Shanghai and to some extent Yokohama, and like the colonial ports of India, was being grafted onto an existing port, and thus it was not a blank canvas of the sort that formed the starting point for mercantile cities planted in the so-called New World. Also, it is essential to remember that Kōbe was a treaty port for a mere thirty years and was an appendage of a larger urban place – Hiōgo. So we must be careful not to overestimate its impact on Japan's development or on the East Asian network of mercantile cities more generally. This study is largely an effort to understand the dynamics of place making under the conditions imposed by an anomalous setting, and to explore how Western colonial forces responded to what was a far more constrained set of liberties than had been the rule in other parts of East Asia. To the extent that a small chapter of a larger drama is *mise en scène,* an account of Kōbe may help us develop a richer picture of Japan's emergence. It seems more likely, however, that much more will be learned about the motives and behaviour of Europeans and Americans during this period – about the colonial mentality, the impact of which suggests that these people were bit players in the drama. In the next chapter, we delineate Kōbe's place in that drama.

Chapter Two

The Creation of Kōbe's Foreign Concession

In 1865 the old port city of Hiōgo was a well-established trading centre. Situated near the eastern end of the Inland Sea, it was close to both Kyōto, the imperial city, and Ōsaka, the second largest city in Japan and the country's principal industrial centre. Hiōgo was situated on a broad, sweeping bay and had a deepwater harbour that offered a measure of shelter, protected as it was by the steeply rising Rokkō Mountains on the north and west and by the Shikoku and Awaji Islands. In Japanese medieval times, Hiōgo had been an important trading centre with links to China. The city also lay astride the Saigoku High Road, the major coast road leading to the west, along which the western *daimyōs* (feudal lords) and their considerable entourages necessarily passed in order to perform their obligatory periods of residence in Yedo every other year during the Tokugawa period. On the eve of the Western entry into the city, Hiōgo probably had around 10,000 people, many of whom were clustered in villages along the shore and toward the foot of the mountains.[1] In addition to coastal trading and the usual urban functions, the local economy included the cultivation of tobacco as well as the national staple – rice, much of which was processed into sake, the highly regarded alcoholic beverage that by then was an important specialty for the settlement and its environs.

The proposed insertion of foreigners into such a setting presented a particular logistical problem for the Japanese authorities. For both the Japanese and the foreigners, the extraterritorial provisions would work best if the foreigners occupied their own space instead of attempting to integrate with an established urban fabric. The tense political climate in the country and Japanese apprehension and resentment of foreigners suggested that the newcomers had better live separately. But it was also important for foreigners and Japanese to live in close proximity so that trade, and the servicing of it, could run smoothly. The best solution – one that had already been implemented in earlier treaty ports in Japan, China, and India – was to graft a new settlement onto an existing city. Accordingly, the Japanese authorities offered to clear a site to the east of Hiōgo, and in 1865 the Minister of the British Legation, Sir Harry Parkes, paid a visit to inspect the location. After spending a number of days there, Parkes declared the site acceptable.[2]

Preparations for the inauguration of the foreign ports of Hiōgo and Ōsaka took place in late December 1867, when Parkes and the consul-designate, Francis G. Myburgh, and other British staff, along with General R.B. van Valkenburgh, the American Minister, and his own consul-designate, Paul Frank, arrived in the harbour aboard their respective naval vessels. Several other British and American naval vessels, as well as assorted merchant vessels, along with representatives of Russia, the Netherlands, and Italy and a throng of merchants, also descended on the place.[3]

The site the Japanese offered for the Foreign Concession was a patch of low, sandy ground, some 1,046 hectares (2,584 acres) in all, about three and half kilometres to the east of the centre of Hiōgo (see Map 2.1).

The foreigners who first viewed the site described the location in somewhat disparaging terms as nothing more than a beach with a few fishermen's shacks.[4] The location did, though, offer access to a very good deepwater harbour presenting none of the navigational hazards that made Tōkyō and Ōsaka ill-suited for foreign trading activities. Closer inspection of the site by the Westerners revealed that although it was sparsely settled, it abutted a small village named Kōbe, whose 1,000 or so residents occupied themselves mainly with fishing and the brewing of sake. Immediately adjacent to the tract, and stretched across it, were a few dwellings and warehouses; nearby were a Shintō shrine and a guardhouse on the High Road leading east to Ōsaka and Kyōto. To the north were rice paddies and a complex of buildings forming the Ikuta Buddhist Temple. The proposed tract was bounded on the east by the Ikuta River, one of the short, flood-prone streams that spilled out of the Rokkō Mountains behind the proposed settlement. This stream ensured a ready source of fresh water both for drinking and for fire protection (the latter was a constant concern of the Japanese).[5] The newcomers were nominally restricted from coming ashore until the day the site was officially conveyed to the foreign powers; however, work had already begun to prepare for trade. For example, the

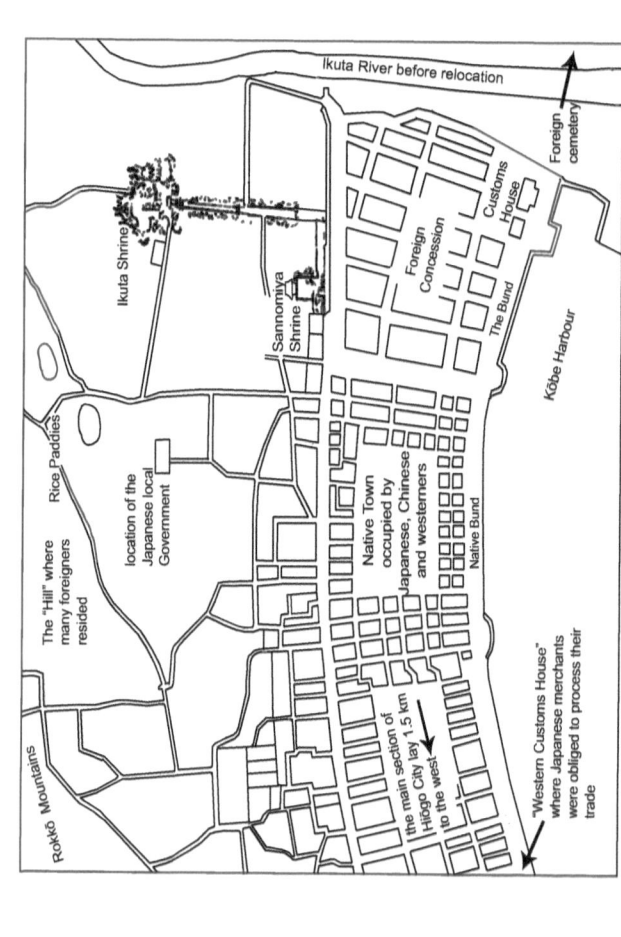

Map 2.1. Map locating the Foreign Concession in its setting east of the City of Hiōgo, ca. 1870. Redrafted by the author from a Japanese pictomap, ca. 1870. Source: Society of Map Archivists, *Collected Maps of Japan's Modern Urban Transformation–Ōsaka, Kyōto, Kōbe, Nara*. Tōkyō, Kaiswa Shobo, 1987.

Japanese authorities, to ensure that the imperial treasury would be able to collect duties once trade officially commenced, had begun constructing a customs house at the southeast corner of the Concession near the shore. Some trade was apparently conducted during the winter and spring of 1868 prior to the completion of this facility; evidently, this activity escaped the collection of duties and taxes.[6]

The foreigners exerted pressure to get the settlement under way. On 3 January, Myburgh reported seven foreign vessels in the port and upwards of one hundred persons of various nationalities already present in anticipation of the opening.[7] The vessels present were mainly of British registry, although an American steamship and a Dutch barque were also in the harbour. Most of these vessels had arrived from Yokohama or Nagasaki, the two most established treaty ports in Japan at that time. It seems that the rising pressure for Kōbe to open had much to do with the poor returns that many of these merchants had been encountering in Yokohama and Nagasaki. Ever hopeful, they were eager to try their luck in Kōbe.

Myburgh also reported that as many as thirty houses in the Native Town[8] had been rented by these persons. Among those who seized this opportunity was George Farley Heard, a member of the American *hong*[9] of Augustine Heard & Co., headquartered in Boston and by then well established in Hong Kong and Shanghai. Heard recorded the acquisition of "a beach lot in the Native Town fronting on the public thoroughfare 40 feet by 116.5 feet, or 104 tsubos for which an annual ground rent of 10 Rios was agreed plus such taxes as shall be in conformity with the Treaty and Convention."[10]

Some contemporary observers of the port's opening painted a vivid and negative picture of this moment. J.W. Gambier wrote that

> a cloud of so-called traders, the scum of the foreign communities from Chinese ports, rushed ashore with indecent haste, bringing with them dozens of cases of whisky, or – as the bluejackets call it, with more force than elegance, rot-gut – with bales of trashy soft goods and all manner of fraudulent rubbish which the unhappy Japanese were to be cajoled into buying; always having to pay the purest gold and silver, at an exorbitant rate of exchange, fixed by the Treaty.
>
> By evening rows and rows of canvas shanties had been built on the Concession ground, whilst a few Japanese houses had been taken possession of and were already in full swing as grog shops. So that, long

before dark, this peaceful, orderly Japanese town, in which for centuries riot had been unknown, was converted into a place where hell seemed to be let loose, with Japanese and Europeans rolling about drunk, fighting, brawling, chasing women, and behaving in a manner which would not have been tolerated anywhere in the civilized world. It was a coarse and degrading debauch this inauguration of Christian civilization.[11]

The foreign powers' eagerness to establish the Foreign Concession was frustrated by the need to carry out basic engineering preparations. Clearly, this was not to be some boom town built in a free-for-all way. Rather, the Foreign Powers envisioned an orderly process of urban planning, one that would reflect their experiences in planting treaty ports on the coast of China and, of course, their earlier attempts in Japan, especially at Yokohama.[12] Much would have to be done to make the site fit for Westerners, and that work would be complicated by the Japanese authorities, who retained ownership of the Concession. In the early months of 1868, an immediate consideration was to ensure a minimal standard of public hygiene by engineering drainage for the tract. Another was to build a sea wall to limit flooding of the site during extremely high tides. This latter project was hampered by storms that destroyed the first several attempts. Much of the land was low lying, and its grade would have to be raised above the high-water mark. This was no easy task, and the drains built on the north and west of the site at first proved inadequate to carry off the water that cascaded down the hills during periods of heavy rainfall. Newspapers in the other ports took delight in commenting on this situation. The *Japan Overland Mail*, referring to the site of Kōbe two months after it was opened, echoed what was probably a widely held sentiment:

> Every day shows us more clearly the disadvantages of the position chosen for the settlement, and within the last few days we have proof that even the enormous sum of money said to have been expended upon it has been literally thrown away. Thirty-six hours of rain has converted one third of it to a swamp, and shows that the fall of land, instead of being towards the sea, leads inland toward the hills which will of course render drainage most difficult and expensive. No one who sees the place can deny that in placing the foreign settlement at Kōbe, the least eligible site has been chosen ... Here according to treaty we are compelled to settle ... Many are of the opinion that the land is not worth the rent they have to pay for it.[13]

Three months later another correspondent wrote:

> Questions are being asked as to why Kōbe should have been selected as the port of Ōsaka instead of the more convenient Sakai ... Kōbe has been weighed and found wanting ... The concession is a swamp and a quicksand knee deep with water in the rainy season, and scorching dusty plain under the summer sun.[14]

Still another wrote:

> The foreign Concession, or the "Sand Patch," "Swamp," or "Desert" as the residents appropriately term it. Remains an incubus upon the hands of the officials and those who planned it.[15]

Finally, one resident, with no small measure of sarcasm and deliberate irony, offered a more hopeful view of the settlement six weeks after the port was opened:

> Sweet Kōbe! Loveliest spot in fare [sic] Japan.
> Who lives in thee indeed's a happy man.
> As through thy pleasant lanes, I take my way.
> Or by the rills pellucid musing stray.
> I think in search of bliss, why further roam.
> Here let me dwell, here make my future home.
> Now with prophetic vision I descry
> A noble city tow'ring toward the sky.
> Palatial godowns meet th'enraptured view.
> Cramm'd with the wealth of India and Peru.
> When Yokohama's crumbled to decay
> And Nagasaki too has had its day.
> This Settlement renowned as Kōbeopolis
> Shall be of Nippon then the proud metropolis.[16]

Far more alarming to those present was an outbreak of armed hostilities between Japanese factions, which so far had failed to resolve the country's political crisis. The British authorities, while they were struggling to prepare the Concession for settlement, received a letter from Hiōgo's governor, Hirobumi Itō,[17] stating that he would be unable to protect the life and property of foreigners resident in the port.[18] Given that only two hundred loyal troops were stationed in outlying posts to the east of Hiōgo at Nishinomiya and Amagasaki, and that the rebels had torched the castle at Ōsaka, Itō's concerns were no doubt genuine. A number of local skirmishes did in fact arise from this turmoil. The most serious of these was on 4 February, when a group of samurai from the Bizen faction fired upon a group of foreigners, including Sir Harry Parkes, the British Minister; that incident, however, was quickly judged by the foreigners to be an isolated one and unlikely to halt their occupation of the site of Kōbe.[19] In the end, the presence of small detachments of British, American, and French troops in the settlement and the inability of the warring factions to master their newly acquired firearms forestalled a serious breach of relations between the Japanese and the newcomers. The business of opening Kōbe thus began.

Before settlement could commence, an orderly survey of the site was necessary, as well as the drawing up of a town plan that delineated streets and lots. A gridiron plan fronting on the harbour was created consisting of twenty-two urban blocks and 126 lots (see Map 2.1). Clearly, the British were taking the lead in opening the port; apparently it was they who prepared the template for the Concession and who executed it on the ground. The earliest surviving Western map of the Concession was prepared by J.W. Hart, a civil engineer, and it was probably he who crafted the town plan itself, given that he was among the earliest residents in the Concession and was later employed by the Municipal Council to design its roads and drains and to supervise the work of building them.[20]

It is not clear what instructions, if any, were provided to Hart, nor what design principles were considered when the plan was being formulated. We are left to deduce those principles from the outcome – not only from the physical delineation of the space itself, but also from the way the urban place evolved after the plans had been made for it. This sort of post hoc, propter hoc deduction is always risky, in that it may attribute motives and outcomes that had not existed at the time of creation. Keeping that in mind, the plan, as noted earlier, involved a gridiron of streets with the waterfront as the baseline. This was to be expected: the gridiron was a standard element in Euro-American town planning of that era. While the rectilinear plan is traceable as far back as ancient Mesopotamia, India, Greece, and Rome, it is not always employed.[21] However, shades of it can be found wherever "controlled" European-inspired settlements have been planned since the late sixteenth century. This is largely because many colonial towns were created on a "blank canvas" from the ground up, or were superimposed on pre-existing settlements that were deemed "erasable" by the

The Creation of Kōbe's Foreign Concession   29

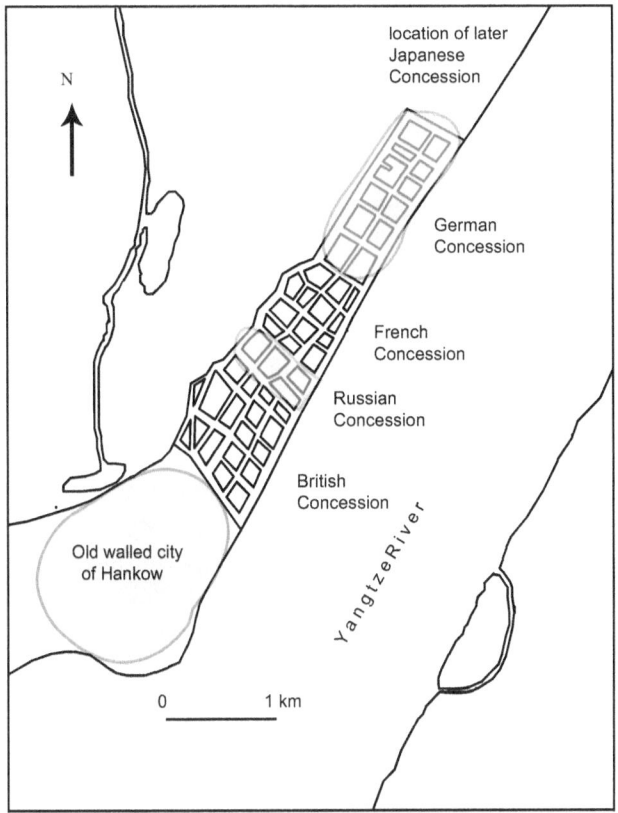

Map 2.2. Map of the Foreign Concessions at Hankow, China, ca. 1880, showing the separate national sectors (based on a map titled "Hankow with Hanyang and Wuchang," in *An Official Guide to Eastern Asia, vol. IV: China* (Tokyo: Imperial Japanese Government Railways, 1915).

European colonists. In many of the latter cases, however – for example, in Calcutta and Madras – the sheer weight of the existing native population forestalled this transformation. Note that many British planners were drawn from the military and imposed the gridiron plan on colonial ports and outposts where they were responsible for defence.[22]

In Japan, which was among the last theatres of colonial penetration, town planners followed the gridiron on the basis of past experience.

It is significant that the grid pattern had been employed for the ports established most immediately before Kōbe – that is, Nagasaki and Yokohama, both of which were founded in 1857. And those two in turn closely resembled the plan for the British Concession at Hankow, one of China's five treaty ports. That one, located inland on the Yangtze River, was opened in 1862 and consisted of separate British, French, Russian, German, and eventually Japanese sectors, which were laid out along the river adjacent to the walled Chinese city.[23] The British Concession at Hankow consisted of a five-by-three-block rectilinear grid fronted by a riverside street. The presence of distinct concessions granted to various foreign nations meant that each had to dovetail with its neighbouring concession to ensure a measure of uniformity. This was achieved at Hankow but not at Tientsin, another Chinese treaty port, where over time seven nations held their own settlements, each of which was planned with seeming indifference to its neighbours.

We might well assume that Shanghai, the most proximate and arguably the most important of the Chinese treaty ports, offered the most compelling model for Nagasaki, Yokohama, and Kōbe. Shanghai's settlement plan was based on a grid, but here again, the division of the space into separate British, French, and American sectors produced a less coordinated result. However, a variant of the gridiron is discernible there, particularly in what became a unified British–American sector after 1862.

The plan for Kōbe, like the one for most port towns, recognized that the front or "water street" would have both mercantile and visual significance, so lot sizes were larger along this avenue. The occupants of these lots would thus be able to exploit their proximity to the harbour either functionally, by building warehouses and other economic installations on them, or by erecting architecturally more impressive commercial edifices, or both. Unlike for towns created under other colonial impetus, such as where the East India Company held sway, there were no forts or garrisons represented in the plan for Kōbe. Indeed, few land use specifications were prescribed. The only evident exceptions were provisions for a Public Garden immediately behind the first block of lots adjacent to the Native Town, and for a customs house, which was to be erected at the southeastern corner of the settlement by the Japanese authorities. The Public Garden, a small, rectangular block, might become an ornamental garden in the Euro-American tradition. It is noteworthy that the British authorities at the treaty port of Shanghai created a similar amenity in 1868.[24] Outside the Concession, across the

The Creation of Kōbe's Foreign Concession 31

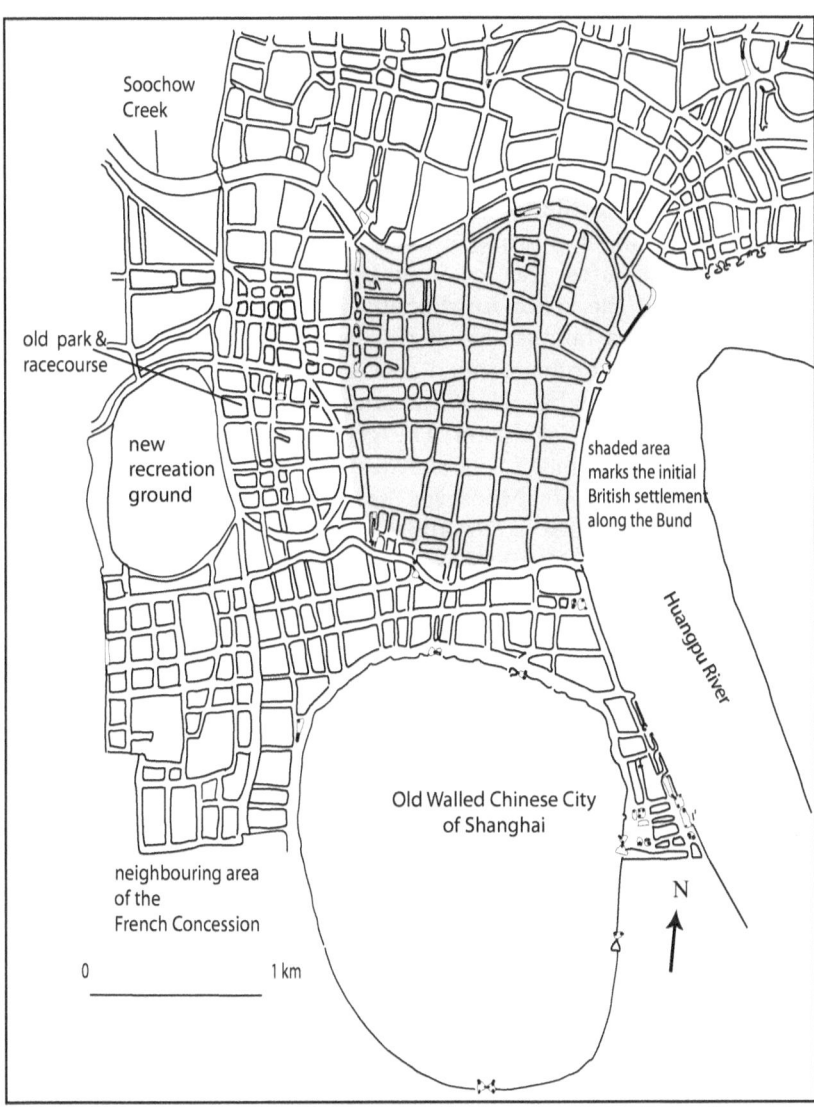

Map 2.3. Shanghai, ca. 1855, showing the British sector of the International Settlement, and particularly the gridiron pattern of the earliest area of settlement.
Source: author, based on *A map of the Ground plan of the foreign settlement at Shanghai, north of the Yang Kang Pang Canal, from a survey by Mr. F.B. Yonel, RN, May 1855.* UK National Archives.

Ikuta River on the east, the Japanese permitted a number of other non-residential land uses as part of the settlement arrangement. In addition to an abattoir, and a hospital for the foreign community, there was a plot to serve as a foreign cemetery, as well as a larger plot to serve as a recreation ground.

There was no single national colonial impetus, no military, religious, and other institutional functions viewed as essential and thus requiring prominent or strategic sites. Rather, the plan seems to have anticipated that after the lots had been auctioned, the occupants would sort out the space functionally in the ways that best suited what was to be a familiar albeit truncated settlement landscape. In this the British and other foreign powers were undoubtedly heeding the lessons of Shanghai, Hankow, and other treaty ports in China as well as those of Yokohama. The grid pattern would ensure that Kōbe would be familiar in its site design and aspect, in its flow of streets and scale of lots. Moreover, its close proximity to an existing albeit alien urban place would allow for some of the economic, institutional, and social functions expected of the mercantile ports with which Westerners were familiar. However, in the absence of stronger evidence pertaining to the planning process, we can only guess whether the result was a skilful and prescient design that foresaw the particular character of the community it would house, or whether what transpired by way of place making was itself a series of independent atomistic responses by a multi-national set of actors working within a simple template.

Closer examination of what unfolded as the Concession was laid out reveals that except for continuing Kaigan-dori (literally: Seafront Avenue) along the harbourfront, the plan made little attempt to accommodate itself to the pattern of streets of what came to be called the "Native Town," which abutted it on the west. Within the Foreign Concession, Kaigan-dori became popularly known as the "Bund," a carryover from its counterparts fronting the Huangpu River in Shanghai and the water streets at Hankow and Yokohama.[25] Initially, Governor Itō had specified that this street should be 100 feet wide – a curious decision, given the prevailing Japanese dimensions for thoroughfares.[26] Later he abandoned this notion, after it became apparent that if this width was implemented and extended into the Native Town, it would be necessary to relocate many of those living along the right of way. Eventually it was concluded that the roadway should be 60 feet or about 18 metres wide. Part of this expanse became a pedestrian "promenade" along the seawall then being constructed along the front of the Foreign Concession. As the plan for the Foreign Concession unfolded, apart from Kyo-machi, the main ceremonial street running perpendicular to Kaigan-dori and north from the harbour, most street widths were set at 10 metres or 33 feet, a decision that was presumably made by the foreigners rather than the Japanese authorities.[27]

Lots varied in size owing to irregularities in the grid. As already noted, the lots on Kaigan-dori facing the harbour were purposely larger. Such "double lots" – Yokohama had them as well – were seen as the most prestigious for foreign *hongs* and other trading and commission agencies by virtue of their waterfront location and the possibilities they allowed for erecting buildings with a stronger identity and architectural pretensions. Most of the remaining inland blocks consisted of six back-to-back lots, each lot fronting a north–south street. In general terms, lots closer to the rear of the Concession were smaller. That said, the typical lot was a rectangle of 900 to 1000 square metres or 10,000 square feet – not large if one hoped to develop business premises that integrated offices, housing, and facilities for storing and processing tea, silk, or imported goods.

Those waiting to establish themselves in the Concession were frustrated by the slow progress of site preparation and by the uncertainties regarding how the lots would be made available. In March 1868 the would-be occupants met and drafted a memorandum to be sent to the ministers of the Foreign Powers in Japan.[28] Specifically, they sought the following information:[29]

- Whether the site at present known as the Foreign Concession would finally be accepted.
- When (i.e., on what date) the site would be ready to receive foreigners.
- What state the site would be in when taken over.
- Whether there would be an "upset price or minimal starting price" on the lots, and if so, what it would be per *tsubo*.
- Whether a jetty and related facilities would be constructed.
- Whether they would be able to obtain property free of Japanese interference.

Ten days later, in a letter to the consuls, regarding the last of these queries, the Governor of Hiōgo assured foreigners that they were free to enter into agreements with Japanese citizens to purchase land and remove houses and that these properties would be liable for Japanese tax. Access to land inside the Concession, however, would be governed by the arrangements contained in the Convention of May 1867.[30]

It appears that as the time drew near to auction the lots within the Concession, several of the foreigners already in the port sought to delay the event. At a public meeting in late August, those present resolved by a 13 to 12 vote to delay the auction. Mr Edward Lowden, who had taken over as British Consul following Myburgh's untimely death, reported that the meeting had been contrived and that many who had voted carried no legitimacy because they were clerks and juniors in firms and could not therefore purchase lots themselves. Lowden deduced that a few people who had acquired water frontage in the Native Town were trying to prevent the sale, no doubt because their locations would lose their primacy, or because they wanted to inflate the resale prices of their own holdings.[31] Evidently this faction failed to delay the sale, which took place as scheduled.

The auction, participation in which was limited to foreigners whose governments had treaties with Japan, was held on 10 September 1868 in the new customs house, which had been constructed in the Western style. The event generated moderate interest, although for some lots there was spirited bidding. All but ten of the lots numbered #1 to #45 were sold (see Map 2.4). Only one lot besides these was claimed. The prices ranged from $0.61 to $1.43 per square metre, with the average price being about $0.75 per square metre (11 square feet).[32]

The Japanese retained the right to levy an annual "ground rent," which was set initially at $0.33 per *tsubo*, as a form of property tax. This was to be collected and tendered by an elected Municipal Council established to administer the Concession and ensure its good order. In addition, each landholder was levied one-third of a *boo* (3 *boos* = $1.00), or approximately $0.10 per *tsubo*, payable to the Municipal Council as a police rate. This meant that the owner of a large lot was liable for annual taxes of about $100 per year, a reasonable amount given the access to trade and favourable lifestyle the port afforded.[33]

Most of those who purchased lots were agents of the *hongs* or trading houses that were staking out a place for their respective firms in the port. The British Consul, Lowden, reported that eleven of the buyers were British subjects.[34] Among them were representatives of Aspinall, Cornes & Co., a British firm already established in Yokohama, which purchased lot #1 on the Bund. This firm served as agents for a prominent British shipping line, the Peninsular and Oriental Steam Navigation Company (the P & O), and its prominent position on the Bund must

The Creation of Kōbe's Foreign Concession   35

Map 2.4. Sale of lots in the Hiōgo Foreign Concession, by date of auction.
Source: Author, reconstructed from information in *Kōbe kaikō sanjnenūshi* and the *Hiogo News*, 18 May 1870.

have been viewed as of strategic importance. Lot #2 was purchased by the American firm Walsh & Company, based in New York, which was already well established in both Nagasaki and Yokohama and had initially set up operations in the Native Town under the direction of their agent Arthur Gay. Farther along the Bund was Smith, Baker & Co., another New York commission agent, which purchased lots #3 and #4 (more or less equivalent in area to lots #1 and #2). This firm served as agents for the Pacific Mail Steamship Company, an American line. Lot #5 was bought by the Netherlands Trading Society. Other European firms and agents were also well represented. For example, lots #8, #9, and #10 were purchased by Schultze, Reis & Co., Textor & Co., and Gutschow & Co. respectively. While the evidence does not allow us to link these firms to specific European cities, they were probably based in Hamburg or Bremen and seem to have been relatively small commission agencies. Farther inland in the Concession, cheaper lots apparently went to individuals aiming to establish service businesses or who had worked for trading firms in Yokohama or Nagasaki and were using the opening of Kōbe to start out on their own. For example, E.H. Hunter, who purchased lot #29, had worked for the firm of E.C. Kirby in Yokohama; St John Browne, who purchased lot #26, had also come from Yokohama.[35]

There were subsequent land sales in June 1869 and May 1870, so that within two years of the port's opening, all the lots in the Concession had been alienated.[36] Judging from the brisk pace at which land was taken up, trading firms and individuals were eager to stake out Kōbe's ground for commercial profit. That many of these individuals had links to other ports in Japan and East Asia suggests a pattern of fluidity; they were testing their luck from one port to the next, seeking to establish a more secure base for realizing their economic ambitions. A few among them were probably land speculators from neighbouring ports, viz. Shanghai, Yokohama, and Nagasaki, who later regretted their purchases when they did not see much future for Kōbe. Not surprisingly, a number of lots soon changed hands. Some of these resales were conducted by lottery, a method that aimed for quick results, if possible. Indeed, lotteries became a craze in 1869 in Kōbe, the prizes being lots of land, a velocipede, and a music box, all of which – except for the town lots themselves – were hard-to-get imported novelties that probably found a ready market among the Japanese.[37]

Not everyone bent on locating in the port chose the Concession as the place to conduct business. Initially there were no port facilities fronting the Concession, whereas those services existed farther to the west in the port of Hiōgo. Those foreigners who sought to service mariners – as mechanics and blacksmiths, as ship's chandlers, and as keepers of sailor's hostels and grog shops – found that the Native Town was a more logical and apparently more available location to set up business. But the Native Town also came to serve as an initial base for a large number of early arrivals simply because the Concession was still under construction.

Many of these new arrivals continued to operate from the Native Town for some time. Indeed, an analysis of advertisements in the local English-language newspapers indicates that at the end of 1869, there were thirty-seven businesses located in the Native Town and forty-one within the Concession itself. A cursory review of these businesses suggests that the more prominent and well-capitalized trading firms, banks, and hoteliers were by then moving rapidly to erect buildings in the Concession that would support their initial requirements. Businesses in the Native Town tended to have less capital; for them, renting temporary premises there was a necessary step prior to establishing in the Concession.

There was also a pattern of foreigners occupying living quarters beyond the boundaries of the Concession. Initially, many traders and their employees chose this option because there were no dwellings within the Concession itself while it was being prepared. Later, the option of living on "the Hill" – the rising suburban ground inland from the Concession (see Map 2.1) – attracted many who presumably believed it offered a healthier environment, particularly in the steamy summer months. Initially, these foreigners rented from Japanese who owned dwellings there, and the Japanese authorities tolerated this in order to expedite commerce. Under the treaty, those who lived outside the Concession were subject to Japanese law with respect to landlord–tenant relationships and local property taxes. Many who had leased or purchased property outside the Concession soon grew worried that without extraterritorial protection, they would be subject to the vagaries of a legal system they did not understand and within which they were at a decided disadvantage. Evidently some felt they were at risk of arbitrary action by their landlords, an anxiety that must have been heightened by the unsettled political climate that prevailed early in 1868. Those foreigners who found themselves outside the Concession sought the help of the foreign consuls to ensure that their living arrangements would be secure even though outside the protections offered by the Concession.[38]

It appears that the Japanese governor sought to reassure these occupants that they would not be subject to arbitrary actions; nevertheless, the foreign consuls sought in the fall of 1868[39] to have the Japanese local government agree to the following:

1. construct properly drained roads and bridges, keeping them from being damaged during flooding of the paddy fields,
2. tax land owned by foreigners outside the Hiōgo settlement and within the boundaries of the Kioto Convention of 26 March 1868,

(i.e., between the Ikuta and Uji Rivers and from the Hills to the beach) at a rate of 10 yen per year per [parcel] of 300 tsubos,
3. every five years there shall be a review aimed at reducing rents and taxes; in no case are these to be increased,
4. that there will not be any interference in the lease terms of land or houses sublet by Japanese to foreigners unless by mutual consent,
5. absolute title deeds already granted will not be included in this agreement unless by consent aimed at providing uniformity for which compensation from local authorities should be offered,
6. this arrangement shall in no way interfere with any other term of the Convention.

These proposals were clearly disingenuous and self-serving, especially the one that sought to ensure that taxes and rents would never rise. Not surprisingly, the Japanese authorities rejected them out of hand. Even so, the Hill quickly became a highly coveted address for foreign residents. The rising ground with its views over the settlement and harbour, and the mixture of farmland and exotic Japanese rural dwellings, offered an attractive alternative to the more crowded and tightly controlled townscape of the Concession. Another attraction of this "suburb" was its distance from Euro-American social conventions; one might more easily keep a Eurasian or Japanese mistress, or share accommodation with male colleagues and partners.

Constructing a New Landscape

In the months following the auction, many lot owners began building permanent premises. Work apparently proceeded with dazzling speed. Since most of the newcomers had initially solved the problem of personal housing by renting existing houses either in the Native Town or on the Hill behind the Concession, the most immediate need was to construct business facilities. For the *hong* merchants this meant a warehouse where consignments of trade goods could be stored when they arrived from Europe or America and where goods purchased locally could be prepared and readied for shipment home to metropolitan markets. The solution was a one- or two-storey building known as a "godown." That term, derived from the Malay word *gudang*,[40] was by then part of the lexicon of the Far Eastern trading world and referred to any warehouse. Five months after the land auction, the *Hiōgo News*

The Creation of Kōbe's Foreign Concession   39

reported the following buildings under construction along the Bund within the Concession:[41]

Lot 1	Aspinall, Cornes & Co.	erecting a very large godown
3&4	Smith Baker	beginning a brick godown
6	Adrian & Co.	erecting an iron godown
8	Schultze Reis & Co.	erecting two godowns & framework for a residence
9	Textor & Co.	erecting a brick godown & magnificent dwelling
10	Gutschow & Co.	erecting a stone godown and a dwelling in front
12	Kniffler & Co.	erecting a brick godown and dwelling house

Three months later, the newspaper could also report:

5&19	Netherlands Trading Soc'y	erecting a godown
18	Warren Tillson & Co.	erecting store and dwelling house
20	Dr. Schokker Hunnink	erecting a neat bungalow residence
25	Semiche, Faber & Co.	erecting a godown and residence

The godowns erected in Kōbe were carefully constructed to be fireproof as well as secure against theft. The *Hiōgo and Ōsaka Herald* described their features as follows:

> As slight shocks of earthquakes sometimes occur, the style in which the Japanese erect their own godowns is generally adopted by foreigners and consists of an interior strengthening framework of timber, around the outside of which and clenched to which with iron clamps the outer walls of stone, brick, or clay are raised. The roofs are peculiarly massive, as over the rafters first a course of boards is laid, then a course of shingles, then courses of clay from 3 to 4 inches thick, embedded in the final course of which the earthen tiles are placed. Shutters and doors are sometimes of iron plate, but more frequently of clay with a hard stucco finish which is again, if on the exterior of the building, protected from the weather by tin or copper plates. Usually in addition to the outer door or shutter hung on iron hinges, there is an interior door or shutter of clay sliding in grooves, to cover the opening … Some of the new erections in the concession are however of brick, without the inside supporting framework mentioned

Figure 2.1. The wide expanse of Kyo-machi during the 1870s, looking northeast. The line of imposing buildings on the right includes the Hong Kong and Shanghai Bank (with colonnade) and the Oriental Hotel (with double veranda). In the background is the smokestack for the Kōbe Paper manufacturing plant.
Source: Harold S. Williams Collection at the National Library of Australia, ACT Australia.

above greatly decreasing the risk of fire. The report argues that the [insurance] premium of 3% per annum is too high and should be reduced to 1% because of the quality of the building and the settlements' amenities.[42]

Brick-walled godowns were preferred, and pan tile was of course the preferred local roofing. To guard against intruders, iron grills on windows and large ironclad doors and locks were desirable. Undoubtedly the godowns built for Augustine Heard & Co. were typical. The Heard premises in the Native Town consisted of a pair of small two-storey warehouses, each 17 by 15 feet, separated by a storeroom 13 by 6 feet. This structure formed a perimeter along one street. Along the opposite side of the lot was a larger godown 38 by 27 feet, which helped define the outer edge of the property along two sides. Within the garden walls that completed the compound was a two-storey dwelling of approximately 40 by 35 feet, which, in addition to providing ample living quarters for the agent, included space for the agent's office and another for that of the *banto* (clerk). This house had verandas on two sides that overlooked the gardens, well, and yard.[43]

Walled enclosures became the convention in Kōbe; many who built there, however, made sure that their buildings presented a public face to the street. Indeed, those businesses located on the Bund and along Kyo-machi Street soon erected facades of remarkable architectural detail, typically with expansive two-storied open galleries overlooking the harbour. Meanwhile, a variety of other buildings were being constructed in the Concession. These included hotels, retail businesses, municipal and institutional buildings, and, of course, houses for those taking up residence in the town. In later years, after the promenade along the sea wall had been lined with lawns and maturing planted trees, the Bund took on a look of self-conscious commercial importance, expressed through a carefully honed aesthetic. Thus, the harbourfront at Kōbe looked nothing like a crude but functional dockland. Moreover, in every respect it stood in marked contrast to the Japanese urban landscape that lay a few hundred metres to the west. Visitors to Kōbe would comment repeatedly on this sense of order, which helped earn the port a reputation for being a cut above many of the other Western beachheads in the Far East.

Chapter Three

Establishing Municipal Government and Services in the Concession

The agreement for establishing the treaty port at Hiōgo anticipated that the settlement would be governed at the municipal level by a council. By 1868, those taking up residence in Japanese ports expected this sort of provision. However, the experience of other earlier settlements was that these governance arrangements left a great deal to be desired. Contemporary English-language newspaper editors generally agreed that Shanghai's municipal administrative arrangement worked reasonably well within the British settlement. By contrast, Yokohama, the most important treaty port in Japan, and the one closest to Hiōgo, was anything but a model to be emulated. The settlement at Yokohama suffered from having been created *by* the Japanese authorities *for* the foreigners, and while a Municipal Council system was later established there, it failed to generate a sense of ownership among the occupants.[1] By 1868, Yokohama had developed a reputation for rancorous internal affairs and intractable municipal government. No doubt because of this, Hiōgo and Ōsaka, as the late arrivals, took pains to chart a better course. Among the critical determinants of success would be the composition of the council and its freedom to exercise leadership in the creation of a functional urban entity.

Because both settlements had been established primarily to support trade, the power balance between traders and consuls seems to have been especially important. There was often a yawning social gulf between the career consuls and the traders. Several of the former had been recruited from the aristocracy or were a part of the rising professional bureaucracies of the European nations. By contrast, some of the traders had come out of the rough-and-tumble world of opium smuggling and gun running on the China coast, and as a consequence, their style and instincts clashed with those of the consular community. Undoubtedly, many of the problems that confronted places like Yokohama reflected the clashing expectations and aspirations of these leading lights. Yet it was also the case that the treaty ports were by their very nature beset by political and structural contradictions that challenged any attempt at local governance.

Chief among the challenges to be confronted was the severely truncated legal status of these settlements. The treaty ports were alien entities embedded in the sovereign political space of a host nation, whose sufferance had to be recognized and acknowledged, regularly and with diplomacy. These obligations were often difficult to meet when the Western community perceived itself as morally and culturally superior and having therefore an unfettered right of occupation. Reinforcing the ambiguity of the Western presence was the fact that the Concession's entire land base was being held under a lease arrangement with the Japanese. From the outset, this altered many of the familiar European and American assumptions and legal conventions associated with freehold land title, the formation of capital from land, the right of electoral franchise, and so on. All the while, however, under the provisions of the several bilateral treaties between the Western Powers and the Japanese, both civil and criminal legal matters rested with the consuls.

This fragmentation of legal jurisdictions left the Municipal Council with anything but clear or obvious means to resolve legal matters that arose within its jurisdiction. For example, a resident living within the Concession who wished to resolve a legal dispute with another resident of the Concession had to seek out his own nation's consul and ask that he resolve the issue with the consul of the other party to the dispute. When the dispute was between a resident of the Concession and a Japanese party, who necessarily lived outside the Concession, resolution became more complicated, for the consul had to reach an accord with the Japanese authorities. This posed the challenge of finding common ground between two quite different modes of jurisprudence. For those who occupied business premises or houses outside the Concession, legal matters pertaining to their affairs were complicated by the implicit assumption that they were subject in the first instance to Japanese legal proceedings. In such cases the consuls had to act delicately to find a resolution that did not threaten the Japanese sense of sovereignty.

This absence of a single unified legal recourse for settling individual intra-Concession and extra-Concession property disputes was troubling. The resolution of legal disputes over land leases outside the Concession invariably required the parties to resort to Japanese courts, where Westerners immediately found themselves at a disadvantage owing to language barriers and weak knowledge of legal precedents and custom. At best, then, an individual living in the Concession might attempt to solve his legal and political difficulties by persuading his consul to take up a dispute or cause with one or more of the other consuls. But because some of the consuls were *ex officio* members of the council, some of these disputes and issues would from time to time impinge on the council's discussions, such that the matter became a public issue. Individual or collective appeals made to the consuls and thence to the council might ultimately be referred to the Council of Foreign Ministers in Tōkyō, which acted as a kind of supra-jurisdictional body for interpreting treaties and for resolving difficulties with the Japanese authorities at the national level.

Figure 3.1 delineates the complex four-part hierarchical structure of political and legal authority that mediated the affairs of the foreigners attached to the Foreign Concession at Kōbe. The Municipal Council derived its role and purpose from the arrangement worked out between the Japanese and the ministers when the port was opened. The arrangements for Hiōgo and Ōsaka differed in their details from those set out earlier; Sir Harry Parkes, the British Minister, had been adamant that the mistakes made at Yokohama not be repeated. Principal among the provisions worked out by Parkes was that the council would have some measure of fiscal responsibility and flexibility to enable the provision of basic services, public works, and communal amenities.

This arrangement was achieved through a formula derived from the initial land distribution within the concession. Specifically, the auction of land leases was configured so that the "upset price" for land was to be eight *boos* per *tsubo*, with six *boos* of this to be tendered to the Japanese authorities directly in recognition of their granting and preparation of the site with respect to constructing and maintaining seawalls, river channels, and landing places.[2] The remaining two *boos* per *tsubo* were to flow into a municipal fund, which might be used to construct public works and other municipal amenities deemed necessary by the foreign community. Furthermore, the Japanese agreed to relinquish to the municipal fund a "moiety," or approximately half of all monies realized above the upset price. As a further measure, there was provision for the council to collect an annual ground rent on each lot amounting to one *boo* per *tsubo*. A portion of this – viz. 1,641 *boos*,

Figure 3.1. A schematic chart of the political structures used for interpreting treaty provisions relating to municipal governance from 1868 to 1899.

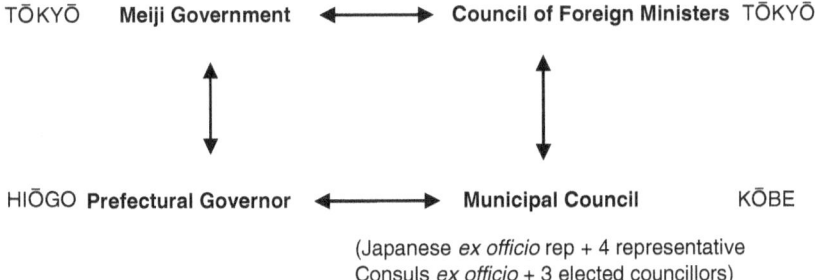

(Japanese *ex officio* rep + 4 representative Consuls *ex officio* + 3 elected councillors)

Source: Author.

or $547 – was owed annually to the Japanese authorities in lieu of a land tax, and the remainder was to flow to the municipal fund for the council's use. Finally, it was understood that if the Municipal Council decided that a foreign police force was required, it would have the right to levy each land renter an annual police rate of one-third of a *boo* per *tsubo*. At the outset, of course, the problem facing the municipality was the significant initial cost of basic public works within the Concession, for which the council was responsible. Funds to accomplish this were believed adequate for the administration of the municipality. Accordingly, the sale of land in 1869 yielded a total of 165,057.31 *boos* or about $55,019, of which approximately $17,665 was turned over to the Municipal Council.[3]

The second provision that Parkes negotiated was that the Municipal Council would have a measure of popular representation. In Hiōgo's case it was proposed that the council be composed of the following: a representative of the Japanese government in the person of the local governor, or his delegate; representatives of the consuls of the treaty nations; and three representatives elected from among the community's registered foreign residents. It was this latter provision that set Hiōgo apart from Yokohama, in that it sought to balance the perspective of the consuls with that of representatives elected by the town's residents.

Behind the scenes among the Foreign Powers there was a difference of opinion as to how these councillors ought to be elected. The British favoured extending the franchise to all "registered foreign residents" of the Concession; however, only those who owned lots would be eligible to hold elected office.⁴ The Americans, on the other hand, favoured restricting both the right to hold office and the right to vote to those holding land. Among the cadre of consuls, there was concern that with so much money involved, a little democracy was dangerous. It was finally agreed that all males registered with a consul would be permitted to vote whether they owned land or lived in the Concession or not, and the first election was conducted on that basis. Nevertheless, some of the Council of Ministers still rejected the breadth of the franchise, and it was only because those elected were "peculiarly qualified" for office that they finally acquiesced.⁵ These patrician arguments aside, what undoubtedly carried the day in this matter was the fact that when the first election was held for council, virtually no one lived in the Concession, since it had just been thrown open for sale. In any event, the seventy-four men casting ballots in the election sent to council K.R. Mackenzie (British, 29 votes), J.S. Blydenburgh (American, 27 votes), and Paul Heinemann (Prussian, 24 votes).⁶ These men won out over sixteen other candidates, which suggests that the first election was a lively one. In addition to the trio of elected representatives, the council included, *ex officio*, the consuls representing Britain, the United States, Prussia, and France. Thus while the council produced only a partial form of representative governance, it did offer the opportunity to weave together a diverse collection of nationalities and interests, although the members seemed on this occasion – as they subsequently would be more often than not – to have been drawn overwhelmingly from the merchant trader class.

As soon as the sale of land was completed and the first election concluded, a newly installed Municipal Council met on 20 October and attempted to focus on the task of ensuring that public roads and other essentials were established to support a well-ordered settlement. But no sooner had it come into being than the Concession's occupants began appearing before it with their problems. One of the earliest meetings was called at the request of the owners of lots #9, #10, and #12, who argued that there were problems with the plan for the street levels and municipal drainage. Evidently Mr Hart's plan called for the roadway to be raised to a higher level than existed on these lots when the purchasers received them. This meant that the drains would have to be raised along with the pitch designed to maintain a flow of water through them. Unfortunately, some people had rushed to build godowns and other buildings on their lots and were now discovering that their properties would be below the general grade of the drainage system.

Discussion proceeded on the advisability of alternative plans. Hart warned that a pumping station would be needed if the drains were sunk lower and that this would be very costly. Mr Blydenburgh, the American elected representative and a man who would quickly reveal an instinct for resisting any expenditure on public works, thought the drains were unnecessary. Mr Evers, the Prussian Consul, and one whose property would be at risk if the change were implemented, threatened legal action, although in the absence of a court with jurisdiction to hear the case, his threat seemed somewhat hollow. In the end the meeting voted 7–2 to proceed with the engineering work as planned by Hart.[7] In a related expression of need, the editor of the *Hiōgo News*, who regularly reported on the Municipal Council, called for a reservoir like the one serving the rice paddies north of the settlement, to be constructed to trap rainwater. This, he argued, could be used to flush the drains every fortnight, during which procedure men would be positioned to sweep the bottoms of the drains. This could be done at the cost of losing a rice paddy or two.[8] At the root of these concerns was fear of cholera or typhus during the coming summer – a fear that was acute among many in the port based on their experience in Shanghai in 1862 and 1863.[9] The consequences of a few occupants having rushed to build on the front lots before the drainage system was in order continued to be an issue in subsequent meetings. Some of these problems became more evident in the winter months of 1869, when parts of the settlement, including the main street of the adjoining Native Town, became a sea of mud, or a "sea of desponds," as the editor of the *Hiōgo and Ōsaka Herald* ruefully called it. In the end, Hart agreed that to avoid having to raise the grade, the drains along the Bund would be sunk lower and covered. Soon after, the contract to carry out this project was awarded to a Japanese contractor named Kisaburo. In what would prove to be a stroke of anticipatory genius, this contract stipulated that there would be performance penalties if the work was not completed on time. Although the motives for this provision can only be conjecture, this decision does hint that several old hands on the council were aware of the cultural difficulties that arose when contracts were concluded with the Japanese. The foreign community viewed agreements like these in the context of Western legal obligations; Japanese legal systems and customary understandings of such agreements were still coming to terms with Western perspectives. The council took further steps to enhance this section of the town when it voted funds to plant trees along the harbour.[10]

These projects proceeded as planned, and as the work approached completion in June 1869, it was decided to extend the covered drains to the second street back from the waterfront.[11] Subsequently, Hart reported that good progress was evident on the sewers; 695 feet had been installed reaching as far north as authorized, and it was expected that the eastern sewer would soon be completed. Drainpipes had been laid nearly the whole length of the Bund, making 2,700 feet in all, and 1,400 curbstones had been laid.[12] However, it soon became evident that there were problems with the contractor, who was judged to be extravagant in the use of materials. Hart wanted a forceful letter on the matter sent to Nakashima, the Japanese Vice-Governor of Hiōgo. By August this dispute had escalated into a significant problem. The council complained in a letter to Nakashima that the contractor, Kisaburo, had failed to complete the work contracted to him and had in fact disappeared. The Japanese authorities, for their part, alleged that the council had failed to pay coolies hired to work on the project and were as a body at fault. It is clear that there were real problems of communication and contractual understanding and that the Japanese authorities were unaware of the written agreement. Mr Crutchley, the editor of the *Hiōgo and Ōsaka Herald*, accused the vice-governor of flagrantly protecting Kisaburo and implied that he was in league with the contractor to exploit the council. The editor further asserted that the consuls had been reluctant to press the issue.[13] Here Crutchley ran afoul of the existing municipal system, which up until then he had stoutly defended, and on whose council sat a largely appointed *ex officio* plurality. In a fit of intemperate frustration, he urged the land renters to pressure the consuls to impeach Nakashima from the Municipal Council, where he had the right to sit as an *ex officio* member, and to press the issue with the Council of Ministers. The council was more sanguine when they met on 12 August to again consider the contract and Nakashima's lack of support for their position. Worrying that they might delay completion of the work if they referred the matter to the ministers, they agreed to accept the proposal from Nakashima under protest. They also specified that they would pay the coolies working on the project at the end of each day.[14] While the work proceeded, the council would appeal to the ministers for an investigation. The irritation continued into September, when Nakashima agreed to meet with a committee to examine the disputed claims over the contractor Kisaburo. On the day of the meeting, Nakashima absented himself because of rain and illness – a move that the council took to be more evasion.[15]

From this episode, it is evident that in the early years of the port, relations between the Japanese and the foreigners were tense, largely because of ignorance of each other's cultural assumptions and patterns of problem solving. Under the surface, mutually destructive racial stereotypes were also at work. Although the council and some of its members sought pragmatic solutions to these tensions, the treaty port era would be characterized by an uneasy racial divide.

Other proposed projects included street lighting for the settlement and the construction of a Municipal Building to house the council, the police, and other municipal functions. There was also discussion about purchasing a fire engine, founding a hospital, and pressuring the Japanese to erect a bridge over the Ikuta River. With respect to the Municipal Building, editorials complained that the Municipal Council had purchased land for municipal facilities and thereby cost the taxpayers money that should properly be used for streets and drains. The writer felt that the council should have been more responsive to Governor Itō's offer to provide land for municipal purposes.[16]

Other concerns were identified as well. Naturally, the Municipal Council discussed the development of a wharf and other waterfront installations, and so did the Chamber of Commerce that had been formed to serve the combined foreign trade interests of Kōbe and Ōsaka. While membership in these two bodies overlapped considerably, the Municipal Council had perhaps some weight to bring to any commercial discussions that involved Governor Itō and the Japanese authorities. For example, it was pointed out that especially after dusk, it was difficult to hire boats (water taxis) at the single jetty that had been constructed within the Concession. Council's view was that another jetty would have to be constructed at the western end of the Bund; accordingly, it prodded the Japanese authorities to act by suggesting that a boat hire facility be created opposite the U.S. Consulate. It was noted that Naka-machi, the street running parallel with the creek and ending at the proposed jetty, had become the main thoroughfare from Kōbe proper to the beach, strangers generally choosing this route in preference to the lacework of narrow lanes of the Native Town.[17]

As for other community requirements, Hart presented in due course a design for a Municipal Hall and a plan for street lighting. Street lighting, using iron light standards, could be installed for the settlement for an estimated 12,000 *boos* ($4,000). This proposal received the council's approval, as did one that established a volunteer fire brigade.[18] The Municipal Building, which would incorporate a post office, a police station and jail, meetings rooms, a fire engine house, and offices for the engineer and secretary, would cost 6,000 to 7,000 *ryōs* ($1,500–$1,750). This plan was not as easily accepted: the editor of the *Hiōgo and Ōsaka Herald* argued that the building, although very attractive as a design, incorporated many irrelevant features – for example, the British Consulate already provided a post office free of charge. He further argued that the octagonal portion proposed for the post office might be used as a Mechanic's Institute and library.[19] Eventually other concerns were voiced with regard to the raising of a Municipal Building. Henry St John Browne, a merchant and municipal councillor, detailed his objections, arguing for prudence with existing funds, since they were all that the council would get for major capital projects. In support of his position, he emphasized the difference between capital revenue and operating revenues, recapitulating these as follows:

OPERATING ACCOUNT	
Receipts	
land rents from sale of September 1868	boos 14,379.17
police rate [year one]	4,793.06
land rent from sale of June 1869	7,416.00
police rate [year two]	2,472.00
total operating fund from all sources	29,060.23
Expenditure	
government land tax	1,641.00
wages for police	7,200.00
uniforms equip't	510.00
lighting	6,600.00
Engineer's salary $500	1,750.00
fire marshal's salary, say $50/mo	2,100.00
rent at $50/mo	2,100.00
sundries, wages	1,000.00
total	22,891.00
Difference	6,169.23
(This to be used for repairs, and cleaning of streets, scavenging, contingencies)	
CAPITAL ACCOUNT	
revenues	
From sale of land 1868	boos 47,339.82
ditto 1869	17,817.00
estimated value of material for drains	1,600.00
	66,756.82
appropriations made	
roads and drains	3,000.00

fire engine	3,000.00
cemetery fence	600.00
municipal lot and fence	3,400.00
lamp posts	5,400.00
trees	500.00
Engineer's commission	1,500.00
	44,400.00
balance	22,356.82

If nothing else, Browne's accounting of the municipality's financial state provides a remarkable summary of what had been accomplished within a year of the settlement's founding. Although trading objectives were slow to achieve, much progress had been made in laying out a town plan, releasing properties to would-be residents, and developing the template for basic civic services in the form of site drainage and public hygiene, road surfaces, street lighting, firefighting, and policing. The public face of the municipality was well on the way to realization in the form of a Municipal Hall; and in the flurry of private construction of houses, commercial offices, stores, and godowns, the Foreign Concession had acquired a neat and progressive appearance, one that was immediately obvious along the Bund, where the wide avenue and sea wall were being dressed with newly planted trees. Few residents or visitors could have been blind to the sense of progress that enveloped the place.

But within a couple of years of the construction of the Municipal Hall, concerns were expressed with regard to Kōbe's system of municipal government. A newspaper report in 1870 referred to a report or petition from the "land renters" that the present form of municipal government was inadequate. The members of the committee of renters included Messrs Gay, Clay, Browne, Mourilyan, Schut, Iveson, Waters, Faber, Holme, Warren, Bessier, Eaton, Korthals, Herhausen, Bovenschen, Joseph, Simon, Look, Scott, Walsh, Behncke, Blackmore, Bonger, Goldman, Richter, Bennett, Vandervlies, Goldsmith, Groos, Grosser, Marmalestein, and Avril – a broad cross-section of the town's merchants and small business owners.[20] Evidently the problem was one of representation and powers of taxation. The petitioners argued that power was vested too heavily with consular officers and merchant landholders, who enjoyed a measure of administrative leverage by virtue of the elected representatives from their class. They asserted that land renters needed to be similarly represented. Proponents of reform petitioned the Council of Ministers to ask the Japanese government for a charter that would assign to land renters the entire control of municipal revenues, as well as the power to levy taxes and frame municipal laws. In effect, they wished to remove the consuls from the council.[21]

It is clear that some of the problems had been festering since 1868 and were rooted in a variety of highly individual land problems. For example, an assertion published in the *Nich Nichi Shinbun*, a Japanese newspaper, alluded to a dispute regarding lots on the Hill.[22] Evidently, early in the opening of the settlement, some foreigners had purchased lots for private residences away from their businesses. At that time, under Japanese custom, the land tax had been paid in kind, and while some foreigners accepted perpetual leases in which the tax was specified in *kokus* of rice, others asked for rent translated into a fixed sum of money instead.[23] The local authorities thereupon named a sum so exorbitantly disproportionate to the tax paid by Japanese holding similar properties, or even to the tax on foreigners who paid in kind, that a dispute naturally arose. After negotiations it was agreed that a temporary lease arrangement would be adopted until a final solution was achieved or permanent title could be arranged. After twenty years the issue still had not been resolved; by then, some of the lots had changed hands, returning to Japanese in some cases, and only nine or ten lots seemed to be still in the hands of foreigners.[24] A solution was arrived at that would calculate the average rent due on equivalent lots under the conventional rental agreement, and this was seen as a reasonable approach. However, a further stipulation was attached whereby the taxes would be assessed in proportion to the size of the lot even though this had not been a provision of other earlier leases. The consuls, anxious to resolve the issue, apparently accepted this stipulation. It also appears there were further provisions that a tenant could not dispose of his property without the approval of the both the Japanese and the consul. The British Consul accepted this provision; the American Consul did not, arguing that it was unjust, given the freedom that other lessees enjoyed. Particularly vexing was that their own consuls had the power to divest them of their property.[25]

Other problems pointed to the considerable difficulty inherent in harmonizing conflicting national administrative styles and political ideologies. In July 1870 the local press, which had hitherto enjoyed access to the council's meetings, were excluded on the motion of the British and Prussian Consuls.[26] The American Consul had vigorously opposed the motion and reacted to its passing by withdrawing his participation on the council. Furthermore, he refused to turn over the rents and taxes collected from American nationals, and he advised

American residents to disregard anything the council did. The three elected members of the council also withdrew, including the treasurer, who took all the accounts with him. This dispute remained unresolved for more than eighteen months, during which time minutes of meetings were provided to the press even though their attendance continued to be denied. With the council's affairs essentially at a standstill, the British and the American Ministers ordered their consuls to restore the council to a functional level. Moreover, the British Minister expressed concern that the consuls' domination of council business was leading to extravagant use of the funds available.[27] Finally, in February 1872, the council rescinded the offending motion and the American Consul again took his seat and paid the rents and taxes collected.

Tensions remained, however, and it was not long before the British and the Americans again found a means to vent their mutual mistrust. In the summer of 1872 the Municipal Council decided to combine the posts of Municipal Superintendent and Municipal Engineer, with the new post to carry an annual salary of $3,000. Accordingly, the post was advertised, and filled by Mr Herman Trotzig. No sooner was this done than the American *chargé d'affaires* intervened, alleging that the press had been excluded from the meeting during which this decision had been taken. He further argued that two of those who had participated were not consuls at all and that Trotzig was, by his own admission, incapable of performing all of the duties required. Others defended the two men who had acted as consuls, arguing that they were locally recognized as filling that function and thus had acted appropriately. It was further asserted that it was understood that the press might attend meetings, but that decisions regarding appointments would be conducted *in camera*. Finally, it was argued that Trotzig was already fulfilling the duties of the post and was receiving a salary.[28]

By the end of 1872 a new system of elections had been introduced in which each elected member was given a term of three years but elections were staggered so that one member would be elected each year.[29] This measure did nothing to resolve the imbalance between the consuls and the elected members; from this point on, however, consular behaviour would be such as to increasingly leave all financial matters to be decided by the elected members.[30] While this may have come to pass because of behind-the-scenes conventions dictated by the ministers to their consuls, it may equally reflect a stage in the evolution of the municipality itself. Simply put, the significant decisions with respect to the public works required to establish the Concession having been completed by 1872, the business of the Municipal Council seemingly became rather more routine and uncontentious. The early 1870s proved to be a period of disappointing trade, and some of the community's energy was lost as high rates of turnover in the resident populations became evident. In 1876, however, when a new phase of public works was launched – this one involving the conversion to gas for street lighting – the emotions over municipal affairs again rose to the boil. The election of January 1876 had been vigorously contested, with voters hearing allegations that the costs of that conversion had been excessive.[31] Animating the election was the fact that one candidate was chairman of the Hiōgo Gas Company. Some argued that he was running for office in order to cover the true facts about the expenditure. It was also argued that the electoral register was hopelessly out of date. No sooner had the election taken place than one council member resigned.

Allegations that the council had overspent on gas lighting appear to have had a basis in fact. The capital account had had assets of over $28,000 in 1873; this had shrunk to $2,000 by 1875, and the anticipated cost of the gas lights would have expended even that remaining sum. As a consequence, the council faced going into debt to meet its salary obligations.[32] As it turned out, most of the capital had gone not to gas lighting but rather to cover the cost of constructing the Municipal Building and jail. It also seems that the community's appetite for low annual taxes had played a part. The council had decided in 1875 that it would be easier to pay the costs of policing out of the municipal fund than to chase residents for the annual police levy.[33]

Among the ministerial overseers of the treaty port of Kōbe and its early citizens, there was a clear presumption that the community, through the force of its Municipal Council and by voluntary cooperative means, would provide its citizens with basic services such as policing and firefighting as well as a hospital facility. For Westerners in the second half of the nineteenth century, these services were axiomatic for public order and communal well-being. Next we document the creation of these services and the struggle to sustain them. This is explored against the backdrop of the treaty port's exceptional municipal charter.

Maintaining Order in the Settlement

In the spring of 1869, the Municipal Council's attention turned to other matters such as the need to establish a police force for the Concession. Such services were considered vital to order, given that large numbers of foreign sailors were present in the port. Most of the problems with sailors were related to alcohol and involved brawling and bar fighting. There were, however, more worrying pathologies; apparently, what the editor of the *Hiōgo News* called an infestation of "unemployed and loafing foreigners" (i.e., sailors) had resorted to robbing and beating local residents, presumably to gain the wherewithal to spend in the grog shops.[34]

It was not just European residents of the port who were threatened in this way: Japanese citizens were often targeted by drunken sailors. On at least one occasion, a shot had been fired at a Japanese by one of them.[35] Incidents such as these made it urgent for a police force to be created along with courts to prosecute the troublemakers. The need to combat crime was a continuing editorial theme for several weeks in the spring of 1870; after a case of attempted arson, the editor of the *Hiōgo News* called for a police force to be founded specifically for the settlement itself.[36]

Another perceived problem was the growing community of Chinese, many of whom were associated with the commercial functions of the Concession. Furthermore, the foreigners harboured suspicions with regard to not only their Japanese employees, but also to the few dissident *rōnin* who were still marauding.[37] The early presence of Chinese in the ports of Hiōgo and Ōsaka reflected the need for Western traders to employ them in commerce, largely because the Chinese could more easily communicate with the Japanese by virtue of a shared albeit somewhat differently evolved writing system. They were also familiar with Western processes for conducting the import – export trade by virtue of their experience with the trading houses on the China coast. In addition, some of the export goods leaving the treaty ports were destined for markets in China, so it made sense to have employees who understood that trade. Those trade items included marine products such as shark's fins, seaweed, and dried fish. Nevertheless, the presence of Chinese in Japan was unsettling for the Japanese as well as for the foreigners; both groups harboured their own respective senses of racial superiority.

Exacerbating all of this, elements of the Chinese population quickly began doing business on their own, outside the control of the Western traders. As early as August 1868 the British Consul at Hiōgo, Mr Lowden, reported that forty-seven Chinese were said to reside in Hiōgo; thirty-seven of them were in foreign employ, ten were shopkeepers.[38] He further alleged that the latter were engaging in trade and that their customs house affairs were being conducted by other Chinese in foreign employ, even though they were supposed to be selling only goods they had made or that had been bought locally.[39] Eighteen months later, the British Consul reported that there were now more than five hundred Chinese in the city, "I am sorry to say!"[40]

By April 1869, three foreign police officers had been hired; it was intended that six Japanese officers would be hired in due course.[41] Subsequently it was determined that the estimated cost of maintaining the police force would be 7,200 *boos* per year, or 600 *boos* ($200) per month. This was the amount initially budgeted, but when it was discovered that this exceeded the amount raised from the tax rate, it was determined that the council would need to seek a voluntary subscription.[42] One solution for controlling costs proposed by the editor of the *Hiōgo and Ōsaka Herald* was to employ Malays, as was the practice in many ports in South Asia.[43] Attention also turned to the passing of police regulations. It was decided that regulations would be in English and, furthermore, that English would be the language of the Police Board.

There was also concern about a clause in the regulations that permitted police to leave the Concession to act on the complaints of foreign residents living outside the concession. The council maintained that the community's interests would be served by such a provision, even though it might offend the sensitivities of the Japanese authorities.[44] Many of the foreigners initially resided in the Native Town, and almost all of the sailors sojourned beyond the limits of the Concession; these areas were patrolled primarily by the police administered by the Japanese themselves. This jurisdictional dilemma exposed foreigners living outside the Concession to the perceived perils of Japanese justice. For the council, this warranted a policy measure that extended their own police jurisdiction into the Native Town. This situation generated a good deal of tension in the first months of the settlement as both police forces worked to develop a relationship that would enable the expectations of the two communities and the several national interests to be met.

One problem was what to do with those apprehended. The plans for a Municipal Building included a jail, but that building would not be constructed until 1876. Anxieties ran high in the middle of 1869, and there were strident calls for additional manpower – specifically, for three additional uniformed Europeans to be appointed, and for twelve Japanese police officers to be recruited.[45]

Court jurisdiction was also a major concern for the foreigners taking up residence in the Concession. The treaties, under the principle of extraterritoriality, made it clear that foreign nationals would be subject to the laws and judicial systems of their home countries. The essential corollary to this was that foreigners would not be exposed to Japanese courts and jurisprudence. Following the precedents established at ports such as Shanghai, the foreign consuls in each port would create the court processes to serve their own nationals. However, the sailor population was a diverse one during the second half of the nineteenth century. Thus, while sailors from the "treaty countries" such as Great Britain, the United States, Germany, France, and Italy had recourse to their national consuls in Hiōgo, those from countries such as Russia, the Scandinavian countries, Poland, Greece, the Pacific Islands, Peru, and Chile, among others, did not. When sailors from these latter countries created petty nuisance problems in ports like Hiōgo, prosecution and resolution posed real difficulties. One solution, advocated by the editor of the *Hiōgo and Ōsaka Herald*, was to establish a combined consular court that might act, by virtue of its combined authority, to handle cases where no consul existed or where no treaty provision had been worked out by a foreign power, as was the case with respect to the growing number of Chinese residing in the port.[46]

Very early on, the foreign residents expressed deep concern over the threat to property theft posed by the Japanese. Several residents claimed that small household items routinely went missing, and Japanese coolies and houseboys were suspected. In some cases, these claims took on greater credence when the employee suspected of being the thief went missing. In other cases, traders reported break-ins at their godowns.[47] Such was the case with a robbery at Kniffler & Co.'s godown in the Concession in February 1869. A revolver and manufactured goods valued at $1,500 were stolen. The newspaper reported that a watchman on duty was either an accomplice or was asleep. Suspicion fell on three *sendōs* or boatmen employed by Kniffler, who had disappeared.[48] In such cases, the foreigners had to call upon the Japanese authorities to find and arrest the alleged criminals. If the editor of the *Hiōgo News* is to be believed, there was a rash of this type of crime in the first six months of 1870.[49]

The perception that the native population represented a threat to property may or may not have been justified. It would not be surprising if temptation overcame those who came in contact with what was for them an array of unfamiliar and fascinating goods and devices. It is perhaps significant that theft invariably involved goods rather than cash. It has also been noted that among the Japanese people one cultural manifestation of the Meiji period emanating from their contact with the West was widespread rejection of things Japanese and a corresponding taking up of Western goods.[50] Ironically this sometimes involved the discarding of ancient cultural artefacts and the elevation of trivial Western items such as the tin can to places of reverence and importance. In this context the opportunity to acquire the commonplaces of the merchant's household may well have presented an overwhelming temptation. But it is also likely that the Westerner's view of the "native" as untrustworthy and given to theft reflected a fundamental suspicion of non-Europeans in general. Although many of the diplomatic cadre appreciated that the Japanese were a culturally advanced people, with institutions and patterns of behaviour and practice equal to those in the West, this intellectual appraisal was far from universally held by the commercial class who occupied the ports. Attitudes among that class were shaped by their longer experience in East Asia and by the prevailing values of mid-Victorian Britain and of the other Euro-American societies from which they were drawn. Not surprisingly, then, strong efforts were exerted to ensure that the Chinese, on whom the traders depended, were kept under control.

The Japanese themselves were concerned about this issue, and Governor Itō had declared that Chinese nationals employed by foreigners must register with the consul of their employer by 18 January 1869.[51] This requirement did not go far enough for the editor of the *Hiōgo and Ōsaka Herald*, who called for "entirely clearing the settlement of these pests."[52] At issue was the fact that Itō's registration measures did not guarantee that all Chinese in the port would be accounted for. Not to be outdone, the editor of the *Hiōgo News* proposed a mandatory registration system for all Chinese in the port such that failure to register at the Eastern Customs House would result in imprisonment. This scheme also proposed that Chinese merchants be charged $15 per year for the right to conduct business; for others, a charge of $10 per year would be levied for *compradors*, and $7 per year for other employees, including servants. The proponents of this plan believed it proper to deduct this tax from the wages of those employed by foreign firms.[53] This system was not imposed, and the less draconian system for registering Chinese in the port seems to have lapsed after a time. Nevertheless there were efforts to restore it in 1881. At the instigation of the Police Commission, the Municipal Council formulated a plan for a "servant's register" that would detail "full and accurate particulars of their parentage, birthplace, etc." in the belief that this measure "will prove a most desirable incentive to the honesty of native servants."[54]

Westerners' racial assessments of, and attitudes toward, the Japanese were complex, as was to be expected given that they were the minority in a what they perceived to be a desirable host country. Their attempts to bridge the cultural divide that confronted them produced a wide spectrum of challenges and comment. Many of the problems that arose were due to what might be regarded as political differences on the one hand, and differences in economic and legal cultures on the other. Colouring this relationship was the persistent belief among the Japanese that the foreigners were interlopers who had extracted highly favourable terms for themselves both in trade and in the Concession settlements that they were permitted to inhabit. The Municipal Council and the consuls in Hiōgo repeatedly had to deal with the Governor of Hiōgo and his administrators in sorting out the interlocking infrastructure of the Native Town and the Concession, and also with respect to developments within the settlement, over which the local authorities retained some control.

In the early years of the relationship at Hiōgo, these interactions frequently led to petty disagreements that reflected differences in ways of doing things, although they had little to do with policing or public safety. For example, soon after the settlement was occupied, the Japanese authorities provided a crew to periodically clean the drains running through and around the Concession. It became the practice of these workers to spread the sandy sediments over the surfaces of the streets adjacent to the drains. The Municipal Council regarded this practice as a recipe for spreading disease within the settlement, and their sensitivity to it became particularly acute as the hot season descended.[55]

Other irritations related to differences in business practices and assumptions. One example involved the Japanese propensity for creating monopolistic trading arrangements even at the local level for ordinary goods. In 1871 the Japanese authorities permitted the fencing of an old cattle market located to the east of the Concession. (This action blocked the route to the settlement – an irritation of another sort.) The consuls soon learned that the local authorities intended to set up a number of other cattle markets in the vicinity. These markets were understood to be leased to a group of Japanese who in return for this monopoly were paying the government 20 *ryōs* per month plus

0.5 per cent of cattle sales. Farmers bringing their cattle from the country were required to turn their cattle over to these dealers for a flat rate plus 5 per cent for the care and feeding of the animals. By these monopolistic means, the dealers located in the city were able to manipulate the price of beef and keep it artificially high in the view of Western authorities. Since the foreign residents were the principal market for beef, this was an issue of some concern; the Westerners viewed it as contrary to treaty stipulations.[56]

On other occasions, disputes arose over differences in interpreting contractual rights and obligations. The well-developed Euro-American pattern of legally delineated contracts as the basis of trade and commerce was not mirrored in Japanese practice, where understandings were more implied than explicit and were based on relationships built up over time. Not surprisingly, Westerners were often exasperated that they could not negotiate binding agreements, especially when the Japanese courts offered little satisfaction. One example involved a British merchant, P.S. Cabeldu, who brought suit against a Japanese merchant who had defaulted on a contract. Having experienced problems in previous dealings with Japanese merchants, Cabeldu obtained security for the contract by having the merchant put up five godowns and their contents, principally sake-brewing equipment, as collateral. When the merchant did not pay him for the contract, Cabeldu sought leave to seize the property through the Japanese courts. The Japanese attempted to stall this, but eventually Cabeldu succeeded in seizing four of the godowns. Later, the Japanese claimed that the contents of the sake brewery had not been part of the security. Complicating the matter was the apparent fraud committed by a Chinese merchant who had also made a contract based on certain goods on the premises. The case stalled, and after numerous delays by the *saibansho* or summary court, a frustrated Cabeldu, aided by the British Consul, Mr Gower, tried to sell the godowns, but there were no takers. Eventually the matter was referred to the British Legation; six months elapsed, and in late August of 1872, Cabeldu was still awaiting the legation's resolution. In October Cabeldu was permitted to try again to sell the four godowns seized against the Japanese creditor. The Japanese authorities, however, indicated that since the properties lay outside the Concession, any foreigner buying them would have to tear down all the buildings within thirty days. This ruling was protested and disputed by Mr Wilkinson on behalf of the consulate. A frustrated Cabeldu claimed $5,383.68 for the original claim of $4,764, the addition being interest and watchman's wages. The matter was referred again to the ministers.[57] In another incident, Deputy Governor Nakayama, who as we have seen was a man with whom the Westerners seemed to have developed a much more contentious relationship than with his superior Governor Itō, seemed about to hire Mr Hart to do some engineering work on the Ikuta River. After picking Hart's brain about this project, the governor opted not to employ him; nevertheless, he proceeded to use the ideas obtained from him. Feeling that he had been exploited, Hart protested this situation to the British Consul.[58]

In yet another case, the customs authorities confiscated goods from the trading house of Mourilyan, Heiman & Co., which upon further investigation was found innocent of the allegation leading to the seizure. Nevertheless, the customs authorities seemed reluctant to return the goods, and this caused considerable upset among traders.[59] In a similar way, traders found their legitimate activities and presence a basis for perceived mistreatment and harassment. In October 1871 it was alleged that Japanese guards had stopped six boatmen working for Kirby & Co. and beaten them with bamboo rods for no apparent reason. The Japanese spokesman contended that the boatmen had been trespassing and that they had ignored the guards' orders to stop. The boatmen denied this, and upon cross-examination the guards admitted that the boatmen had not been insolent to them, nor had they misrepresented themselves to the guards.[60]

The problem of theft and robbery perpetrated by Japanese against foreign residents apparently continued into the 1880s. The editor of the *Hiōgo News* detailed the lengthy process facing consuls lodging complaints against Japanese on behalf of their nationals. The British Consul, Mr Ashton, apparently handled the correspondence of twenty-eight cases of robbery during 1880. Fourteen of these cases involved charges against a named individual; eleven convictions resulted.[61]

The episodes of petty theft aside, life in the Concession was generally remarkably orderly and free from the rowdiness that plagued many other foreign settlements in East Asia. The controlled release of land, the relatively high price of property (which kept many people out), and the successful creation of a Municipal Council combined to ensure that grog shops and bordellos were not located within the boundaries of the Concession. Such businesses were confined to the Native Town, which, among other liberties, housed a well-known "licensed quarter" or red-light district known as the Fukuhara.[62] That district was conveniently located along the northern edge of the Native Town until it was displaced by the coming of the railway. The proximity of these services relieved the Concession of one of the greater nuisances that had infested the early foreign quarter at Yokohama. Moreover, because Japanese and Chinese employees of foreign firms were not permitted to reside in the Concession, the occupants of the town were able to ensure that Euro-American social values and codes of middle-class behaviour held sway. These conditions made early Kōbe a particularly attractive location for Euro-American women, although it took several years for the female population of the Concession to approach anything like a balance.

The provision of a jail in the Municipal Hall and the sustaining of a police force, however ill-supported by the municipality, provided another clear example of conflict between consular and community interests. The year 1879 saw a spate of escapes from the jail, which clearly needed stronger security. Questions were raised at this point about the jail's capacity to serve as a prison for those sentenced to longer terms. Some felt that the facility was a "lock-up" for short-term holding, not a "hard labour facility," and argued that the community should not have to pay the facility cost of the latter. The consuls, they argued, had a responsibility to provide such facilities since the administration of justice was in their camp. Speaking for the consuls, General Stahel, representing the United States, contended that the consuls were not responsible except to pay the daily maintenance charge for prisoners in their care. His solution was for the municipality's jail to increase the per diem charged; this would produce sufficient revenue to maintain the place. In the end it was moved that $300 be allocated to increase security by means of an iron grate over the yard. Also, it was agreed to increase the per diem to 80 cents per day from 50 cents. The proposal was passed by the Municipal Council.[63] These measures seem to have had a salutary effect, for when the Municipal Superintendent provided his annual report a few years later in 1885, he was able to report that fifty-nine persons had been housed in the jail for a total of 565 days during the year, which had produced a small profit of $189 from the 80 cent per diem rate.[64] In 1897 the superintendent was able to report that there had been no serious thefts or crime, although 161 prisoners had been kept in custody during the period. Nonetheless, during the same year the per diem for keeping a prisoner had risen to $1.40. In that year the police force still comprised one foreign sergeant, two foreign constables, and thirteen Japanese constables. Registration of servants had been maintained.[65]

Forming a Fire Brigade

The threat of loss of life and property from fire was a significant concern for the Japanese, given that their urban settlements were constructed at a high density and almost entirely of wood. They had developed highly efficient community-based fire watches and response procedures. The foreign residents of Kōbe also saw a clear need for a fire brigade, particularly as their economy came to develop a heavy reliance on the tea trade, wherein much of the tea was "fired" in godowns located within the Concession. The fact that the settlement came to consist of compounds made up of a closely proximate dwelling house, office, and godown made anxieties over fire hazards that much more acute. Indeed, one British Consul had tried at the outset to have the ministers establish a regulation that would have prevented the construction of wooden buildings altogether in the Concession.[66] So it was not a surprise that no sooner had the settlement been occupied and the Municipal Council struck than there were calls for the formation of a fire brigade.

In April 1869 the merchants and businesses of the Concession launched a subscription for the purpose of purchasing a fire engine and other firefighting equipment. The record of expenditures from the Municipal Council accounts suggests that an appropriation of $3,000 was made for the purchase of a fire engine; it is unclear, however, whether this sum was the result of the subscription, or was the amount committed from the municipal reserve, or was some combination of the two.[67] Whatever the case, it was anticipated that the fire engine would arrive in late June of that year. Efforts now turned to the formation of a volunteer fire brigade.

In late July 1869 the General Committee of the Chamber of Commerce tabled a report on the settlement at Hiōgo and Ōsaka for the information of fire insurance companies. That document described how the Municipal Council had established a police force, acquired a fire engine, and initiated the formation of a fire brigade. It also described the flows of water through the Concession and how ample water for fighting fires could be ensured, noting that it was possible to dam the settlement's drains temporarily when necessary. The report stressed the absence of fire in the Native Town due in large part to the effectiveness of the native fire corps, pointing out that only one fire – and that one on a foreigner's premises – had occurred since the port was opened. Finally, the report described how foreigners, when building godowns, were adopting Japanese construction methods, which had been developed specifically to resist fire. It was argued that all of these factors, and the advent of the fire brigade, were reasons to set fire insurance rates at 1 per cent of value rather than the 3 per cent that was typical.[68] There was evidence, too, that the council anticipated that when a Municipal Building was eventually constructed, it would include space to house the fire engine. Moreover, it was anticipated that the council would include a fire marshal among its employees. At least one member of the council proposed a salary of $50 per month for this position.[69]

It is difficult to determine how common fires were in the Concession during the early years. The newspapers periodically mentioned them; for example, in December 1869 it reported that the fire engine had performed well.[70] Significantly, the above-mentioned report noted that the fire brigade had not yet been formed, which suggests that even four or five months after the engine had been acquired, the threat of fire had not been sufficiently strong to make this requirement compelling. Subsequently, efforts to recruit a fire brigade were turned over to the Police Committee of the Municipal Council, and by February 1870 the newspaper was able to report that 150 men had joined the brigade. It also described how they had recently jogged through their first muster, which ended in a feast; as with many such bodies in the West, the brigade gave evidence of being as much a social institution as a utilitarian one.[71]

A decade later, the annual report of the fire brigade for 1880 detailed six fires during the year. Four of these appear to have been at businesses in the Concession; two were in the Native Town, including one at what was referred to as a Chinese house. More important was the brigade's recommendation that four additional fire wells be dug: at the northwest corner of lot #64; opposite lot #39; on Yedo-machi near lot #96; and on the Bund near lot #5. The same report requested an additional storey for the engine house to provide sleeping accommodation for coolies, which suggests that the brigade now employed some permanent members.[72] The maintenance of fire wells became the responsibility of the Municipal Supervisor, and his annual reports would refer to this requirement.[73]

More serious fires occurred from time to time, including one in 1898 at the China Export-Import Bank. There, the fire jumped to an adjoining building occupied by Messrs Reynell & Co. The latter's family had to make an escape. The loss, estimated at 60,000 yen, consisted of the dwelling and the bank; the nearby godown survived.[74] Also in 1898 there was a major fire in a cotton warehouse on the Kōbe Pier, outside the Concession, which resulted in losses estimated at about 1 million yen. This event followed a string of reports of fires – most of them house fires – that resulted in at least one death, that of a Mr Eaton. It was widely suspected that these fires had been arson.[75]

Medical Services

When the settlement was being created, the Japanese authorities proposed erecting a small, three-room hospital to serve all foreigners, especially those landing at the settlement with smallpox and other infectious diseases.[76] This structure would be built on the east bank of the Ikuta River, just beyond the Concession and well away from the Japanese settlement. At the same time, the Japanese identified a site near the hospital as the location for an abattoir, adding the stipulation that no cattle could be slaughtered in the Concession after 1 May 1868.

In initiating these measures, the Japanese governor was showing his strong interest in expediting the opening of the port to foreign trade, no doubt because of its potential for enriching his treasury. The creation of a customs house, marine hospital, and abattoir indicate that the governor understood what a trading settlement required in order to function. The decision to locate what the Japanese presumably perceived as the two most "noxious" facilities – the hospital and the slaughterhouse – across the river in the settlement's southeastern quadrant also indicates the exercise of Japanese values and geomancy – a factor that to this day strongly limits the attraction of certain sectors of Japanese cities. The strategic placement of these functions owes much to the fact that those in Japanese society who were responsible for slaughtering cattle and working with leather were by custom an outcaste group – the *Eta* or *Burakumin*. It soon also became evident to the Westerners that the real purpose of the hospital was not to serve foreigners but rather to fulfil two purposes for the Japanese. First, they were justifiably concerned about isolating and controlling the contagious diseases that might arrive with the large transient population of mariners expected to flow through the port. Providing a quarantine hospital was thus a prudent measure on their part and undoubtedly also served the foreign community in this specific role. Second, the Japanese authorities were keen to acquire Western medical knowledge. Such knowledge had been an important part of what had been gained from the Dutch during the

Tokugawa period, and the appetite for this learning was strong. The Japanese authorities hoped that by building hospitals for the foreigners at the emerging treaty ports, *de facto* medical schools might arise.[77]

By the end of 1869, however, it was clear to the foreign community what the Japanese aspirations were and that foreign residents would not be welcome in the Japanese hospital. Faced with this, they set about creating their own facility under the administration of a Board of Directors, which published a schedule of fees. In early January 1869, Myburgh, one of the British officials sent to open the port, reported that a naval hospital was being located in a house, although it was in need of repair. He also indicated that he was working to obtain a doctor's house.[78] Funded by a subscription launched by a group of public-spirited citizens, Kōbe General Hospital (later renamed the International Hospital) opened in June 1869 in a rented house at the southern end of Ikuta-mae.[79] Running the hospital would prove to be a heavy burden on the foreign community, which did not comprise more than a couple of hundred persons, nearly half of whom were tradesmen, shopkeepers, clerks, and the like.[80] Interestingly, the schedule of fees distinguished first-, second-, and third-class patients and included a provision for Japanese servants of foreign residents. The problem of paying for charity cases admitted to this hospital would generate resentment in the community.[81]

What is clear is that with the opening of the port in 1868, a template for an ordered and in many ways familiar set of civic resources was put in place at Kōbe. Although trade started slowly, the port seemed to have attracted a core of newcomers prepared to make the port thrive. For that to happen, however, it would be absolutely crucial to find commodities to trade.

Chapter Four

Forging an Economy: The Basis for Mercantile Trade

To understand the economic motives of the foreigners who developed communities in ports like Kōbe, it is essential to examine the broader context of their presence there. Japan's long, self-imposed seclusion from the rest of the world, which began about 1635, and which was broken only by the arrival of Perry's Black Ships in 1853, had stifled Japan's participation in international trade. While Japan had developed a highly complex intra-regional trading and supply system during this period of isolation, its truncated involvement with its trading neighbours meant that Japanese mercantile firms were ill-prepared to engage the external trading world when the time came for them to do so. Their lack of acquaintance with foreign markets, financing systems, and sources of supply, and with the mores of external trade, resulted in a vacuum into which the foreign traders stepped easily, supported as they were by the "unequal treaties" that favoured Western interests at the expense of Japanese ones. Thus from the outset the foreign trading firms were able to monopolize the import and export trade from their bases in the treaty ports. This chapter begins by examining some of the technical challenges of reconstructing trade and economic activity generally in Japan during this period.

Trade Statistics and the Currency Problem

The assembling of comprehensive and accurate statistics essential for measuring and analysing trade in general, and that of the foreign port of Kōbe in particular, during the second half of the nineteenth century, has posed a significant challenge to researchers. The Japanese do not seem to have compiled these statistics until the mid-1870s, when the apparatus of the new Meiji administration's Ministry of Finance finally developed to the point where it was possible for them to do so. Initially, statistics were collected by the local customs authorities, but these have been judged to be unreliable owing to inconsistencies and changes in the definitions used and also due to the inexperience of those collecting them at source.[1] In 1872, the administration of customs houses in the treaty ports was transferred to the Ministry of Finance and methods for developing trade statistics were standardized. Until that point, one is dependent on the trade statistics produced by surrogates such as the Chamber of Commerce for Hiōgo and Ōsaka and/or the consuls representing the treaty nations in the port. The recording of these statistics was hit and miss. The British and American Consuls had consular officials in the port whose specific task was to provide annual reports to their superiors on the dimensions and state of trade. The consular representatives of other nations probably did the same, but this research has not sought out their reports. Unravelling those statistics that have been located, and achieving a measure of consistency with respect to them, can be a challenge. For example, during the port's first years, most commodities were referenced by their weight, and the practice was to employ the Japanese *picul* (1 *picul* = 133.3 lbs) for this purpose. By 1877 the American Consul, no doubt to better the understanding of readers at home, had switched to quoting most commodities in terms of pound weight. There were also differences in reporting owing to the use of the calendar year in some cases and the fiscal year in others.

Equally vexing was the parallel practice of placing a monetary value on the commodity being traded. The system for valuing trade goods was based on the market price at the port of clearance. This statistic was important as it provided the baseline for the 5 per cent *ad valorem* duty charged under the treaty agreements between the Foreign Powers and the Japanese. Because of this, there was great incentive for merchants, wherever possible, to value articles of trade in ways that best favoured them. This led merchants to undervalue both imports and exports, but particularly imports, at the time of declaration in order to reduce the duties that had to be paid in compliance with the treaties. It is also noteworthy that some imports seem to have entirely escaped being included in official trade statistics. The prime case in point relates to the import of ships purchased by the Bakufu and by the feudal domains in the early years. No doubt, the Japanese authorities viewed these ships as strategic acquisitions and not as trade in the conventional sense. Nor was the outflow of gold and silver included in the

Japanese trade statistics, although it might be cited in those recorded by the Chamber of Commerce or the consular returns. Finally, we can presume that smuggling was taking place and that this activity escaped official recording.[2]

Quite apart from these problems, establishing the value of goods has proved anything but easy owing to the peculiarities of the Japanese monetary system. At the time of Japan's opening, there was a decided lack of standardization in Japanese currencies. Under the Bakufu the various feudal domains were permitted to issue their own paper currency, although the central government was responsible for issuing coinage. These practices led to regional differences in money standards. For those operating near Ōsaka, prices were expressed in terms of the weight of silver, while at Yedo (renamed Tōkyō in 1868) prices were expressed in terms of the weight of gold. Muddying this situation further, the bullion weighting system used by merchants in Yedo was different from the one used in Ōsaka. Furthermore, in order to raise money for its own use, the Bakufu had over time debased the nation's coinage such that its quoted value bore no direct relationship to the metal content of the national coinage.[3] Isolation from international trade had insulated the Japanese from the consequences of their unorthodox approach to currency valuation. However, with the opening of Japan to trade, these internal differences coupled with differences in the weighting and valuation of gold and silver among Western traders led to a large-scale outflow of gold – or what is frequently referred to as "treasure" – as foreign traders sought to profit from these discrepancies.[4] For Westerners the standard for computing the value of trade goods after about 1850 came to be the Mexican silver dollar, a practice that was widely accepted in the China trade. The value of the dollar coin, even though there were several versions of it in circulation, was based entirely on its content of silver bullion.[5] Thus among Westerners, trade statistics compiled in the first decade of activity in Kōbe were expressed using this dollar valuation. However, beginning with the New Currency Act in 1871, the Japanese initiated monetary reforms in an effort to overcome the currency chaos that existed internally and to curb the outflow of gold from the country. Associated with this was a new Imperial Mint established at Ōsaka,[6] and a new convertible paper currency based on a decimal system and using gold as the standard.

As a result of Japan's inexperience with international currency fluctuations, her determination to be independent, and simple bad timing, this system had begun to collapse by the mid-1870s. One reason was that in an effort to promote mercantile trade, the young Meiji government had minted special one yen silver coins to be circulated in certain restricted areas to facilitate transactions among merchants engaged in foreign trade. The theory was that these silver coins would have metallic content such that they could be equated with both the Mexican dollar and the new one yen gold coin. But in 1875, after large silver deposits were discovered in the western United States, the international value of silver began to decline, with the result that the following year, the "international" price ratio of silver to gold became 20:1. In Japan, the government's conversion rate was not adjusted to reflect this international movement, and as a consequence, "cheaper" silver coins were brought into the country and exchanged for gold coins, which were then exported at a profit to those engaged in this practice. Thus the Japanese government had to abandon the gold standard in favour of a silver standard, and it was not until 1897 that it again returned to a gold standard.[7] Nevertheless, official trade valuation statistics compiled after 1871 were quoted in yen. By 1877 the American Consul was reporting trade conducted by American firms in Kōbe using the yen rather than the Mexican dollar as the currency. All of this introduces further confusion in assessing trade statistics.

A further challenge to interpreting trade statistics for Kōbe arises from the tendency to lump the trade of Kōbe and Ōsaka together. Until navigational difficulties were overcome late in the century, trade activity at Ōsaka was negligible while that at Kōbe was robust – indeed, it was growing to the point that by 1870 Kōbe became the second most important foreign trading port in Japan after Yokohama. Stripping out the small level of trade attributable to Ōsaka is all but impossible. However, given that several of the merchants engaged in this activity were represented in Kōbe *and* Ōsaka, it can be argued that the impact on any analysis is unlikely to seriously skew the picture being drawn here.

It is important to note that the import and export trade conducted by foreigners was not restricted to the simple movement of goods in and out between the home nation and Japan. Much of the activity of foreign trading houses involved moving goods coastwise to other ports in Japan and to the China coast. This intra-regional trade was undoubtedly important to the financial health of the foreign firms engaged in it. Initially much of this trade involved firms sending and receiving items between other branches of the same firm located in ports like Yokohama and Nagasaki. Items procured at other treaty ports could then be consolidated and shipped to international markets. One impact of this practice was that it inflated the level of trade that might be statistically attributed to Yokohama, which was the principal beneficiary of this practice. However, care must be exercised to ensure that trade data are constituted in the same way from reporting agency to agency and from period to period. Finally, it is important to recognize that the international trade conducted by foreign firms in Kōbe and other treaty ports might involve the movement of goods to any port beyond Japan. That is to say, British firms, for example, were not by nature or regulation restricted to moving commodities between Japan and Great Britain; traders could move goods freely between origins and destinations in Europe and North America, as well as to other parts of Asia, Australia, Hawaii, South America, and so on. Indeed, much of the trade involved the exchanging of goods between Japan and China.

Opening the Port to Trade

With the opening of Kōbe in 1868, traders and others looked expectantly toward the first season of mercantile commerce. The two prerequisites for success would be the development of shipping services to support the movement of goods and people and the establishment of an operational trading framework. Kōbe was a latecomer to the commercial world of Japan and the Far East; but to its advantage, it straddled an already developing shipping route between Yokohama to the east and Shanghai and other Chinese treaty ports such as Cheefoo and Hong Kong to the west, a route that included Nagasaki. Not surprisingly, after Kōbe opened it became a regular stop on this route for the steamships and mail packets of the American-owned PMS Company and the British-owned P&O Line. PMS had opened for business in the port as early as 24 November 1868, when it employed Alfred Phelps of Smith, Baker & Co., an American trading firm, to act as its agent.[8] Smith, Baker operated from quarters in the Native Town near the "western" customs house while its premises on lots #3 and #4 in the Foreign Concession were being constructed. Within a couple of weeks, another British trading firm, Aspinall Cornes & Co. of Yokohama, had established a branch in Kōbe, where it served as the agent for the P&O Line.[9] Shortly thereafter, Phelps advised the community that PMS would be operating a semi-monthly service from Shanghai to Yokohama on the steamship *Costa Rica* beginning on 6 January 1869. This vessel would depart Kōbe for Yokohama and have a twelve-hour layover in these ports both coming and going. Similarly, Aspinall, Cornes advertised that the P&O's *Azof* would depart Kōbe for Yokohama on 7 January.

Besides moving the mails, these two lines provided a passenger service that soon became an important means for merchants and others to move along the trading axis that was developing through the Chinese and Japanese treaty ports. It is perhaps a measure of the limited scale and fluidity of early Kōbe's community that for a time, the local newspaper recorded the arrival of important passengers using this service.[10] For example, we know that among those on board the *Costa Rica* arriving from Yokohama bound for Nagasaki in the first week of February 1869 were Messrs Hassage, Benck, Townley, Storey, McCailin, Van der Vot, Adair, and Mathie, as well as three ladies, presumably European or American women of like social class as the men cited. Also on board were four Japanese military officers and 152 Japanese passengers in steerage bound for Nagasaki. In addition to the above, we learn that Mr J. Lyons, H.H. Horton, and T.W. Gardiner of the U.S. Navy and five Japanese officers, as well as 252 Japanese and 11 Chinese passengers in steerage, were on board bound for Shanghai.

We will never know whether this listing of passengers was comprehensive and complete, but if it was, the fact that 315 people were travelling in steerage and that 23 were not suggests a remarkable contrast in comfort level. In any event, these shipping services provided unprecedented opportunities for travellers to move quickly, if not always comfortably, within the trading orbit emerging within and beyond Japan. The regular schedules of these steam vessels also enabled traders to move among the local agencies constituting the trading firm and to ensure that information and small consignments of goods flowed quickly and predictably from point to point as needed. This would have been especially useful during the years when Kōbe's mercantile community was just getting established.

From late May to late December during the shipping season of 1869, three PMS ships – the *New York*, the *Costa Rica*, and the *Oregonian* – made thirteen, fifteen, and four calls respectively at Kōbe as part of this expanding service.[11] The P&O's *Cadiz*[12] and *Formosa*,[13] which were smaller and less regular in their calls, collectively visited the port on ten occasions during this period. Besides these vessels, there were a number of other ships that provided charter services for the merchant houses themselves. Those houses may well have been shareholders in the vessels. An example was the *Bahama*, which called at Kōbe fourteen times, each time in the name of the Adrian Trading Company based in Yokohama.[14] Similarly, the *Sakuara*, a British-registered steamship of 625 tons that worked between Nagasaki, Kōbe, and Yokohama during the season, made eight calls at Kōbe. This vessel provided shipping services for three different merchant houses during this period. The record seems to indicate that ten vessels operated in this patterned way insofar as they made at least two and as many as fourteen calls each at Kōbe over the period. There were, of course, other vessels that made calls in the port, but their movements seem not to have been regularly scheduled. Many of these latter ships undoubtedly belonged to independent chartering firms that sold their services to traders. Some had forged ongoing relationships with these houses such that the same ship was under charter to the firm throughout the season.

The port also attracted a large number of other vessels of varied national registry that conducted the "long haul" carriage of goods between Japan and home ports in Europe and the United States or that worked the trade between the China coast and Japan. These were typically sailing vessels rather than steamships. Many were small barques of 200 to 300 tons, and their visits to Kōbe were singular events, unlike the regular visits of the vessels discussed above.[15] In all there were eighty-four such visits from ships registered in Norway, Holland, Austria, Denmark, Russia, Germany, the United States, Britain, and France during the first shipping season of the port of Kōbe. The numbers of these vessels in port at any one time varied with the seasons, but the report in the *Hiōgo News* of 9 October 1869 gives a fair indication of what was a typical merchant fleet presence in the harbour for this early period. There were seventeen merchant vessels in port that week, and because the newspaper adopted the practice of citing the last port of call prior to arrival in Kōbe, we know that ten had arrived from Yokohama, five from Nagasaki, and two from Shanghai. What we do not know from the record is the more distant backward points of origin of these vessels, nor whether they were carrying cargo from these points or collecting cargo for homeward-bound charters. Whatever the case, we can say that the most significant event that affected the larger pattern of shipping in this era would have been the opening of the Suez Canal in November 1869, for it ushered in a new phase of long-haul shipping employing steamships. Perhaps the greatest impact of the canal, however, was that it transformed the movement of passengers between Europe and the Far East.[16]

Finally, several naval vessels – gunboats, larger warships, and naval survey ships – moved in and out of the port with some regularity during this first season. The presence of the survey ships is a reminder that the major powers were diligently collecting strategic information that would serve their national interests, particularly their commercial objectives. For example, Her Britannic Majesty's survey ship the *Sylvia* recorded six visits to the port during the summer of 1869 as she charted the waters of the Inland Sea near Ōsaka and Kōbe. More ominous perhaps was the presence of three American naval frigates, an Italian corvette, and various gunboats representing the United States, Britain, Russia, and France. In all some twenty-seven calls were registered by these ships, a few of which made two or three visits. It must be remembered that this activity reflected that Japan was undergoing upheaval and civil unrest in the final days of the Tokugawa Bakufu, before the Meiji regime achieved ascendance. As was noted earlier, some of this civil unrest focused on the Ōsaka–Kōbe region; so it is not surprising that the Western powers understood the need to show that the newly opened port and its foreign community were amply supported by military force.

Any discussion of shipping raises this question: What role did the Japanese play in the movement of goods as foreign trade began in Japan in general and in Kōbe in particular? Undoubtedly the penetration of foreign traders into Japan exposed the inefficiency of the traditional Japanese shipping sector. Japan's self-seclusion had seriously delayed its progress in shipbuilding, navigation, and the logistics of long-haul trade. The traditional Japanese sailing vessel was very similar to the Chinese junk, and fleets of these vessels in the hands of long-standing regional shipping firms moved goods up and down the coastal shipping lanes that constituted Japan's largely internalized domestic economy. That said, Western sailing vessels were not unknown to the Japanese, owing to the enduring Dutch presence at Nagasaki, and foreign contacts had increased in number and scope during the first half of the nineteenth century as American and Russian whalers and Western traders took refuge from storms in Japanese ports or set out deliberately to test Japan's policy of exclusion.

It must be borne in mind that Western seafaring, too, had been undergoing a remarkable technical transformation since the late 1700s. Oceangoing vessels had become progressively larger and faster and capable of moving greater and greater quantities of goods with ever more reliability. Quite apart from the revolution in ship architecture and the development of dazzlingly complex systems of sails, the scientific knowledge base associated with ocean transport had improved dramatically: maps were more accurate; internationally standardized positional reckoning had been devised; and there was greatly enhanced understanding of weather and climate as well as ocean currents.[17] Also, there had been marked improvements in the management of those who manned long-haul vessels. Finally, the growth of global commerce had resulted in a complex logistical system that provided oceangoing ships with provisions, repairs, and eventually coaling facilities as steam power took over international shipping.

Advances in steam power and in new, more durable materials for constructing ships were rapid by the time Japan opened to the West in 1853. Not surprisingly, there was great interest in these developments at the highest levels in Japan, not only in the Bakufu but also among the dissident clans of Satsuma, Tosa, and Chōshū, all of which were challenging the Shōgunate for power from their bases in western Japan. Given this dynamic, second- and third-hand ships of European design were among the very first commodities acquired by Japanese interests after trade began. Typically, they bought these ships directly from foreign traders rather than through brokers. In this way, the Japanese acquired at inflated prices ships that were already technically outdated and ill-suited to their needs. Many were small paddle-wheelers designed for river or inland lake service rather than for ocean or coastal shipping. These transactions frequently left the purchaser in financial difficulty and encumbered by vessels that failed to meet expectations. By the time of the Meiji Restoration, some 20,000 tons of Western-style shipping capacity had been acquired in this way. The drive to acquire modern ships as rapidly as possible would continue under the new Meiji administration, many of whose leading thinkers were the same men who had led the rush to acquire Western ships as Japan was being forced open.[18]

Much of this interest in shipping was rooted in the perception that Japan needed to become a naval power as quickly as possible in order to establish its place in the world order and to defend its sovereignty in the face of the "semi-colonialism" that the "unequal treaties" represented. As soon as Japan opened in 1859, foreign traders, and the Western shipping lines that accompanied them, poised themselves to dominate its mercantile economy. For Japan to wrest control of even part of this trade from the foreigners, it would have to acquire the means to conduct international trade, and to achieve that objective, mastery of commercial shipping would be necessary. The challenge was formidable: even if Japan could somehow overcome the knowledge deficit, a competitive shipping sector would be very capital intensive. Reform of the existing industry therefore became essential, particularly because by 1870, foreign firms had established a thriving business moving goods along Japan's coastal routes.[19]

Soon after it was installed, the Meiji government issued the Merchant Shipping Rule of 1870, which mandated the transformation (i.e., Westernization) of the Japanese shipping fleet.[20] However, traditional domestic shipping firms were slow to take up Western practices, and they tended to acquire European and American sailing vessels rather than steamships. This reflected a conservative habit of mind, one predicated on step-by-step progress toward acquiring knowledge and technology. After 1870, the fleet of traditional Japanese vessels remained in place and in fact expanded. In 1885, in a further attempt to compel the domestic industry to change, the government prohibited the construction of any ships of the traditional type having a cargo capacity greater than approximately ninety tons.[21]

The domestic shipping industry's reluctance to overhaul its merchant fleet led to the creation of a largely new segment of Japanese industry: the government awarded a select few firms a monopoly on the acquisition of Western ships. It has been argued that most of those who succeeded in exploiting this monopoly were new players in the industry and were not weighed down by the millstone of tradition.[22] This may be true to a point, but the firms behind these endeavours were already well-established entities that as we shall see were led by men who combined valuable connections to the ruling regional (and later national) oligarchies with entrepreneurial flair and a vision of what Japan might become if it could control its own commercial and military destiny. Chief among these entities was Mitsubishi Bussan, led by the remarkable Yatarō Iwasaki (1834–85), a man associated with the Tosa clan, who began his involvement in shipping as part of the maritime defence and commercial development agency established by the Tosa domain.[23] Through that endeavour, Iwasaki gained managerial experience as well as knowledge of the new modes of shipping. Once it had been coupled with the monopolistic system adopted by the Meiji government, the firm would rise to become one of Japan's first international success stories in the shipping sector.

The Meiji government awarded Mitsubishi a monopoly on the transport of military men and materiel at times of crucial national interest such as the Taiwan expedition of 1874 and the Satsuma Rebellion of 1877. This entailed more than a conventional commercial monopoly; the Japanese government went so far as to *give* the firm ships so that it could accomplish the tasks expected of it. Moreover, the favoured status enjoyed by the firm was underpinned by massive subsidies not only for acquiring ships but also, later, for transporting mail. Ultimately, these strategies and the managerial acumen developed by Mitsubishi led that firm to engage the leading Western shipping firms – the Pacific Mail Steamship Company and the P&O Line – in a cut-throat struggle for dominance on the key routes linking Japan to Shanghai and other points on the China coast.[24] That said, throughout the period, foreign traders continued to rely predominantly on Western shipping firms to move cargo to and from home markets. Thus, the record of ships and shipping linked to the foreign port of Kōbe remains largely one involving Euro-American ships and related foreign-owned services (see Figure 4.1).

The pattern evident in Figure 4.1 has three striking features. First, British and American shipping interests were by far the dominant foreign presence in Kōbe throughout the period covered by these data. Second, Japan's domestic shipping industry began to achieve prominence during the last four years covered, once Japanese firms took over the mail packet services formerly conducted under the American flag. To the extent this was so, there was a noticeable decline in the numbers of American ships visiting the port after 1875. A third observation is that trade activity apparently declined between 1874 and 1876. This collapse can be explained by a financial crisis in Japan resulting from the new Meiji government's inexperience with international trade and currency standards. This problem will be discussed below.

It is a measure of the fundamental role of trade in the early life of the port that the consuls and principal agents of the leading trading enterprises formed a Chamber of Commerce almost as early as they did a Municipal Council. Another essential for trade was the creation of shore-based portals for trade in the form of a customs house, bonded warehouses, and mechanisms for hiring stevedores. The Customs Service was a responsibility of the Japanese, and while they were reasonably quick to erect premises for this purpose, they were less able to make them operational or to develop hours of service that suited the Westerners. While this building was under construction in the first months of mercantile activity, officials did not collect duties.[25] As much as traders may have welcomed this situation, they soon found much to complain about.

One of the earliest initiatives of the Chamber of Commerce was to make representations regarding the inefficiencies and costs of boat hire and coolie labour and the insufficiency of the hours of business of the

Figure 4.1. Number and Nationality of Ships Entering The Port of Kōbe, 1868–80.

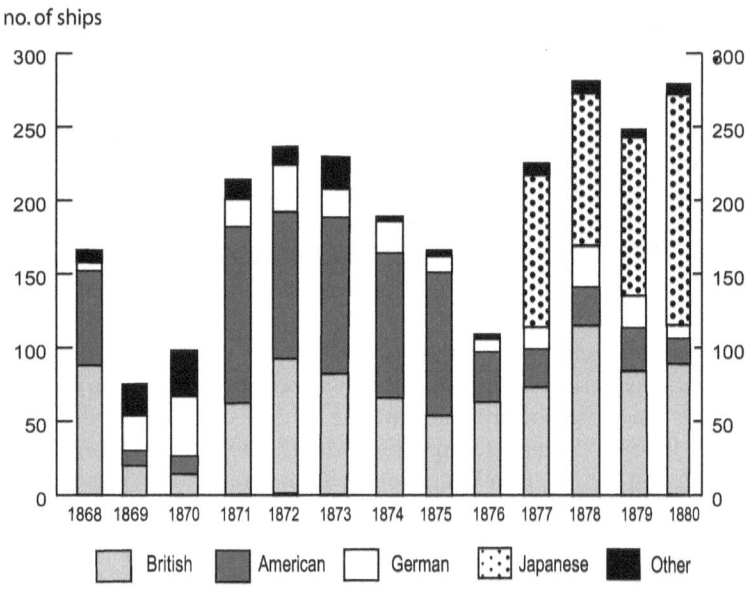

Source: British Parliamentary Papers, "Return of Foreign Shipping Entering the Port of Hiōgo," 1868–80.

customs house.[26] Communication of these concerns to the Japanese authorities rested with the consuls, who later reported that Governor Itō had accepted the concerns and was open to creating better means to facilitate trade.[27] Further exchanges of correspondence indicate that early in the life of the port, there was a shortage of insurable godowns, with the result that traders had to rely on the Japanese government's bonded warehouse for both the landing and the storage of duty-free goods and those on which duty was to be paid. This problem was soon alleviated by the construction of private godowns by the trading firms themselves. No sooner was this accomplished than the Chamber of Commerce sought a reduction in the insurance premiums being charged in Kōbe, arguing that the insurers had failed to recognize the very substantial and generally fireproof qualities of godowns in the port. They further argued that exceptional firefighting capacities existed in both the Native Town and the Concession, owing to the constancy of water flowing in the drains running through these settlements, which when dammed ensured an ample reservoir of water with which to fight fires.[28] While these arguments may sound self-serving, those who made them had a strong case: their construction standards for godowns, including the materials employed, were such that there would be few disastrous fires in spite of the widespread practice of tea firing and the storage of highly flammable commodities such as kerosene in these buildings.

Finding Goods to Trade

The problem of finding goods to trade was a vexing one for those who settled at Kōbe. In the buoyant optimism of the age, it was assumed that trade was always possible. Indeed, by the time Kōbe was opened, the Japanese had been trading with the foreigners for about a decade. Traders were confident therefore that there would be opportunities to sell Western goods in Kōbe and nearby Ōsaka; however, finding Japanese products to balance that trade was initially more problematical, particularly as the tensions and civil unrest surrounding the final days of the Bakufu undoubtedly induced a wariness among local merchants and regional producers of goods. The newly formed Chamber of Commerce quickly began assessing the prospects for trade and determining where impediments might be overcome either through cooperation or by lobbying the Foreign Powers, or the Japanese authorities, or both. Preoccupying the Chamber was the prospect of tea and silk as possible export staples for Kōbe. In October 1868 the Chamber struck a committee to investigate the supplies and quality of these commodities and the appropriateness of the current tariffs on them.[29] The committee reported that silks brought to the Kōbe market differed in quality from those seen in the Yokohama market. The principal classes of silk in the Yokohama region were fetching prices ranging from 1,400 to 1,600 *boos* per *picul* (or $467 to $533 per *picul*). It was reported that toward the end of the previous season there had been quantities of superfine Otsu, Ide, and Gosho silk held in stock at Kyōto and Ōsaka for the use of domestic manufacturers; the high prices paid by foreigners for such quality goods had induced local merchants to offer silks to the Kōbe market at prices averaging $611 per *picul*.[30] The duty of 5 per cent would accordingly add another $30.55 per *picul* to the cost of exporting these goods.

The report went on to discuss the international supply situation resulting from the failure of the silk crop in Europe and the drop in exports from China, such that demand had inflated prices in the home markets. The authors of the report were well aware that market conditions might change; indeed, they expected the market to normalize shortly, and they argued that an increase in duties would be injurious to the market when this happened.

With regard to tea, the authors of the report found it difficult to arrive at any very accurate picture of the average value of the teas brought to market at the port. During the first two or three months a considerable amount of business had been done "at prices which certainly would not have exceeded 60 boos." However, "eager competition amongst buyers has considerably increased the value of this article." The committee considered the price of 60 to 70 *boos* (or $20 to $23 per *picul*) to be a high average, particularly given the 5 per cent duty.[31]

As elsewhere in Japan, there was also the difficulty of identifying which members of the Japanese mercantile community might be willing to conduct trade. Another problem was establishing the protocols and relations of trust under which this activity would proceed; still another was how to valuate currency when money transactions were used. The latter issue bedevilled commerce in all of the treaty ports. By the time that Kōbe opened to trade, the Foreign Powers and the Japanese authorities had been conducting more than a decade of complicated negotiations to arrive at a satisfactory basis for currency valuation.[32] As indicated previously, this period had been marked by episodes of very active currency speculation during which foreigners bought up and removed large amounts of the gold, as well as the copper coinage then in use; the Japanese authorities had responded by further debasing their coinage, thereby triggering a severe inflationary period. These problems were still being resolved in 1868, and the elements of trade shown in Table 4.1 indicate that much of what transpired in the way of trade can best be understood as a response to various aspects of this "crisis." For example, the inflation set loose by the currency problems within Japan had caused the domestic price of rice to increase more than fourfold between 1859 and 1868.

Although the amount of rice moving through the foreign port at Kōbe was small, it does suggest that merchants were responding to opportunities that were available to them within Japan. What is not indicated in the summary of trade reported in the *British Parliamentary*

Table 4.1. Trade at the port of Kōbe, 1868[i]

Imports	From England and other countries	From open ports in Japan	Subtotal
Textiles	$289,801	$2,468,155	
Metals	174,320	204,768	
Arms/ammunition	125,266	1,267,614	
Raw cotton	194,322	33,497	
Raw sugar	1,572	32,512	
Coals	4,666	–	
Saltpetre	10,000	–	
Miscellaneous	11,157	1,089,139	
Imports	**$811,104**	**$5,095,685**	**$5,906,789**

Exports	To England and other countries	To open Japanese ports	Subtotal
Silk	$263,811	$ 528,958	
Silkworm eggs	18,666	146,702	
Cocoons	–	7,767	
Tea	29,776	280,089	
Tobacco	–	164,306	
Camphor	135	125	
Seaweed	6,503	2,431	
Fish	650	–	
Wax	15,727	24,920	
Timber	1,270	692	
Drugs	634	5,250	
Rice	–	168,658	
Miscellaneous	49,709	334,218	
Exports subtotal	**$388,096**	**$1,664,116**	**$2,052,212**
Total overall trade	**$1,199,200**	**$6,759,801**	**$7,959,001**

[i]Even though the data used to construct this table covered trade at both the "foreign ports" of Kōbe and Ōsaka, the trade passing through the latter was so small that these figures largely reflect the movement of goods through Kōbe. That said, many of the imports of cotton and textiles, treasure, and armaments would have been destined for Ōsaka via trans-shipment.
Source: *British Parliamentary Papers*, "Commercial Report on the Port of Hiōgo and Ōsaka," 1868.

Papers is the movement of "treasure" or specie that saw some $948,790 enter through the port and $2,749,106 exported.[33] This, combined with the commodities itemized in Table 4.1, added a further 46 per cent to the gross trade being conducted at the port. Clearly, the traders were attempting to play the speculative markets for gold and other precious minerals within the country. Whether this "treasure" was in the form of old coinage of higher value, or articles made from gold, silver, or copper, which could be employed in place of debased coinage, cannot be conclusively determined from the aggregate data available.

Reliable currency valuation and exchange was a problem for traders, especially when the flow of information was slow and when other actors engaged in the day-to-day activity of resolving exchange. Newspapers during that period often reported on this issue. For example, the editor of the *Hiōgo and Ōsaka Herald* noted that in August 1868 the effective exchange rate was 256 *boos* per $100, a valuation that indicated a substantial depreciation of the dollar, given that the treaty specified an exchange rate of 311 *boos* per $100. Perhaps the explanation for the depreciation of the dollar currency was the common practice of leaving the matter of money payments and exchange in the hands of Chinese *compradors*, who, so it was alleged, "squeezed" a part for themselves in these transactions.[34] Merchants were apparently ignorant by choice of the details of this activity. The editor further claimed that the *shroffs* (clerks) in the Bank in Yokohama were profiting from taking in bad or inferior trade dollars only to pass them out again as good coin to other bank customers.[35] These problems continued through 1869, with the editor reporting that the Japanese program of replacing coinage such that old *boos* were exchanged at a rate of 342 to 343 per $100 was causing the *niboo kin* (the 2 *boo* coin) and new *boos* to fall into disfavour. It was further argued that the value of imported stock on hand, as well as that of export goods, was being depreciated because of the Japanese minting of "under equivalent" coinage. The paper reported that "the unsatisfactory position of exchange of both gold and silver currency continued to exercise a most unprecedented depression on imports, and business was being brought almost to a standstill."[36]

These problems would continue to form a cloud over trade for several years. In 1876 the American Consul at Kōbe complained to his superiors that people had been importing U.S. silver trade dollars from Shanghai and Hong Kong and having the Japanese Mint recoin them as yen. He claimed that during August and September 1876, $530,000 U.S. trade dollars had been recoined in this way, and because the weight of U.S. silver coins was heavier, this practice produced a net profit of 25 *boos* (per $100), or about 7 per cent, to those who engaged in it.[37]

It is clear that the initial year of trade was very unbalanced, with imports of Western goods far exceeding exports of Japanese products. As Table 4.1 shows, the principal imports that year were Western munitions and textiles (principally cotton and woollen goods) as well as the specie to pay for those goods. Almost nothing of like value left the port, and much of what did seemed destined for other points within Japan, where it might be used to supply other domestic markets or be consolidated by the more established trading houses in Yokohama and Nagasaki for movement to China or elsewhere. For this reason these figures present a somewhat ambiguous picture of the real trade being conducted at the ports. This pattern was still evident in 1870, as shown by Table 4.2.

Although the trade was still imbalanced if measured in monetary value, some local goods did begin to be exported. Silk and related products, along with tea, stood out as items finding a market. Eventually, the variety of regional domestic products for sustaining trade would expand to include rice, camphor, vegetable wax, cotton rags, and curios. By far the most important export, however, was Japanese tea. Expectations that tea would be a viable commodity were evident in the 1868 report of the Chamber of Commerce for Hiōgo and Ōsaka:

> With regard to Tea it has been found somewhat difficult to arrive at any very accurate knowledge of the average values of the quantities that find a market at this port. During the first two or three months a considerable amount of business was done in all of its classes, at prices which certainly would not have exceeded Bus 60 [sic]. Latterly, however, an eager competition amongst buyers has considerably increased the value of this article.[38]

Trade thus commenced at Kōbe with muted enthusiasm. Over the first couple of years, those who lacked capital, connections, and experience were induced to move on or to find employment with more established trading firms. Only the hardiest traders were able to weather the choppy period of initiating trade at this location at a time of general economic depression. We now look at trade volumes and change over the succeeding period. Table 4.3 plots a time series of aggregate trade at Kōbe in comparison to its principal rival, Yokohama. It is clear that as the period wore on, Kōbe gradually increased its import and export activities and made slow but progressive inroads with respect to catching up to Yokohama as a significant port. The overall pattern was

Table 4.2. Summary of trade through the ports of Kōbe and Ōsaka, 1870

Imports	From foreign ports	From Japanese ports	Total
Cotton goods	$318,659	$ 658,062	$ 976,721
Woollens	358,116	784,363	1,142,479
Satin	12,912	–	12,915
Metal	25,928	34,516	60,204
Arms/ammunition	91,382	71,294	168,676
"Eastern produce" (rice, sugar, cotton)	2,281,369	239,058	2,520,427
Miscellaneous	689,936	162,367	851,803
Gold and silver	1,350,000	4,680	1,354,680
Subtotal	**$5,133,605**	**$1,954,340**	**$7,087,945**

Exports	To foreign ports	To Japanese ports	Total
Raw silk	$ 427,993	$763,487	$1,191,280
Cocoons	11,210	26,000	26,000
Silkworm eggs	3,150	1,185,000	1,188,450
Tea	796,013	707,507	1,503,520
Miscellaneous	1,161,401	370,809	1,532,208
Subtotal	**$2,399,567**	**$5,452,668**	**$5,452,668**

Source: British Foreign Office, Correspondence from the British Consul at Hiōgo, Japan, Public Record Office, Kew, FO 262/40 28 January 1871.

of growth in trade, with absolute increases in the aggregate value of trade at Kōbe in all but five years (1876, 1881, 1882, 1891, 1899).

It is useful to set this pattern against the ebb and flow of overall business fluctuation as shown in Table 4.4. It is difficult to draw a close correlation between these short-term Japanese cycles and those occurring in Europe and America, except where the cycles seem to reflect the effects of financial liquidity and money supply. Indeed, Tsuru cautions that for the period before 1874, the monumental task of adjusting Japan's social and economic fabric resulted in a sufficiently chaotic pattern that it is difficult to see the Japanese economy being in any way comparable to those in Europe and America. In this sense we might better view the trade activity taking place at Kōbe and Yokohama as reflecting economic forces at work in the overseas markets alone. For this reason an attempt has been made to delineate the economic health of these overseas destinations as the primary force in producing these swings.

Forging an Economy 85

Table 4.3. Imports and exports at the port of Kōbe, in yen, 1873–99

	Imports	as % of imports at Yokohama	Exports	as % of exports at Yokohama	Overall Trade
1873	3,924,974	19.88	2,516,896	16.04	6,441,870
1874	4,769,195	28.48	3,405,534	26.13	8,174,729
1875	5,354,919	23.76	2,852,380	21.99	8,207,299
1876	3,789,669	20.03	3,450,298	15.82	7,239,967
1877	4,257,689	20.25	4,657,180	29.26	8,914,869
1878	6,026,150	24.23	6,505,376	41.86	?2,531,526
1879	7,067,155	29.94	5,750,203	29.86	12,817,358
1880	7,848,537	29.81	5,653,853	29.78	13,502,390
1881	7,380,529	34.19	5,588,804	26.02	12,969,333
1882	6,378,820	31.34	6,514,972	24.20	12,893,792
1883	6,989,189	36.36	5,972,653	22.91	12,961,842
1884	7,790,531	40.03	6,610,733	30.27	14,401,264
1885	7,584,148	39.91	7,255,844	29.95	14,839,992
1886	9,499,172	47.11	9,933,661	31.19	19,432,833
1887	13,851,161	50.97	12,770,606	37.81	26,621,767
1888	24,667,906	67.31	18,304,070	44.96	42,971,976
1889	26,035,331	75.86	20,331,553	48.57	46,366,884
1890	32,041,004	78.83	16,955,413	52.44	48,996,417
1891	25,700,501	88.67	21,733,718	43.87	47,434,217
1892	30,698,177	97.99	21,295,740	34.60	51,993,917
1893	41,294,276	113.74	24,968,974	45.23	66,263,250
1894	56,910,503	112.81	29,438,113	40.32	86,348,616
1895	63,098,427	112.48	38,307,955	45.18	101,406,382
1896	82,546,593	113.38	40,317,817	65.35	122,864,410
1897	110,741,831	127.53	51,408,080	56.68	162,149,911
1898	138,133,798	124.43	60,119,645	74.86	198,253,443
1899	120,289,525	157.34	75,320,884	69.56	195,610,409

Source: Based on statistics published in *Nihon boeki seiran* (Foreign Trade of Japan – a Statistical Survey) (Tōkyō: Toyo Keizai Shimposa, 1975).

Not surprisingly, the fluctuations in trade correspond to those seen generally in both the Japanese and the global economy throughout the period. Perhaps noteworthy, however, is that the downturns were comparatively modest in absolute monetary terms, and that several of the upturns represented very substantial increases that surely gave encouragement to traders in the port. For example, during the five years from 1883 to 1888 the gross value of trade at the port rose by more than 230 per cent, and from 1889 to 1899, it increased by more than 320 per cent. Significantly, Kōbe was steadily catching up to its

Table 4.4. Economic fluctuations in Japan, 1868–97,[i] with reference to broader external business cycles

1868–70	Depression (related to collapse of a period of gold speculation in United States)
1870–74	Upswing
1874–76	Downswing (probably related to money shortage in United States in 1873–8 and start of the so-called Long Depression ending in 1896)
1876–81	Upswing (resuscitation of U.S. agricultural and railway building)
1881–85	Downswing
1885–90	Upswing
1890–93	Downswing (correlates with rise of silver price via the U.S. Sherman Act, which triggered a rise in Japanese currency)
1893–97	Upswing (running counter to the Panic of 1893 in United States)

[i]The business cycles shown here were produced by *The Oriental Economist* in 1931 and form the basis for Tsuru's analysis of the forces at work behind them.
Source: Shigeto Tsuru, "Economic Fluctuations in Japan, 1868–1893," *Review of Economic Statistics* 23 (1941): 176.

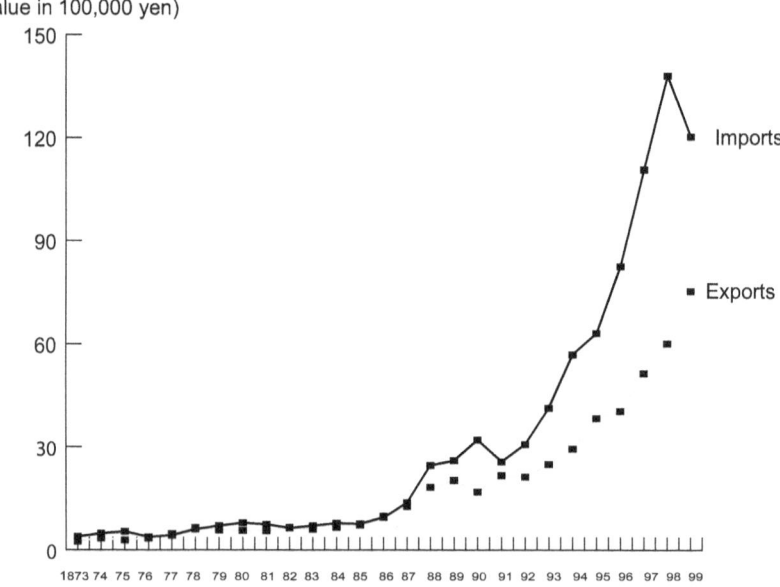

Figure 4.2. Growth in the value of trade at Kobe, 1873–99

Source: Based on statistics published in *Nihon boeki seiran* (Foreign Trade of Japan Statistical Survey) (Tōkyō: Toyo Keizai Shimposa, 1975).

Forging an Economy 87

rival, Yokohama, during the period. While it never overtook Yokohama in the export trade, Kōbe did surpass its rival in imports in 1893 and retained a healthy superiority in this component thereafter. The explanation as to why imports were so strong at Kōbe lies in its role as the recipient of large volumes of cotton destined for the textile mills of Ōsaka. We turn now to a fuller exploration of the components of trade, particularly those that came to sustain Kōbe.

Chapter Five

Finding a Mercantile Staple for Kōbe: The Tea and Silk Trades

The first merchants to locate in Kōbe were concerned with establishing a toehold in the port in order to sell various Western goods to the Japanese. These goods included used munitions and ships, which the Japanese had shown a disposition to purchase with specie. Only later did it become essential to find local domestic products to sustain trade. In the older treaty port of Yokohama, the two items that had emerged to fill this need were tea and silk, and Kōbe's early merchants recognized that these products could, for them, become important and perhaps sustaining export staples. Many merchants commenced preparations and investments in anticipation of establishing these items as the core of Kōbe's trade. This chapter reconstructs the local supply and dimensions of trade in both these commodities. We start by looking at the silk trade.

An Overview of the International and Japanese Silk Trade

Silk production, called sericulture, involves harvesting the filament forming the cocoon of the silkworm *Bombyx mori*. The silkworm has four stages in its life cycle: the egg, the silkworm, the pupa, and the moth. The silkworm feeds on mulberry leaves and forms a covering around itself by secreting a protein-like substance through its head. Producers harvest this filament under tension and wind it onto reels. The threads may be wound together to form yarn. After drying, the raw silk is packed according to quality. This process was developed in ancient China and dates to about 3000 BC, if not earlier. Sericulture reached Japan through Korea around AD 300 and then travelled west to India; it was known in Rome by the AD 500. During the eighteenth and nineteenth centuries, Europeans made several major advances in silk production. These innovations included improved silk-weaving looms, power looms, and roller printing.

The opening of Japan coincided with a moment when the silk market in Europe faced a supply crisis owing to a disease that had ravaged the production of raw silk in France and Italy.[1] This was also a time when the demand for silk was expanding in the West, particularly in America after 1870. As a result, silk formed a substantial part of Japan's total exports throughout the period of this study, reaching 46 per cent of gross trade in the 1870s. This fell to 35 per cent in the 1890s; however, in terms of absolute value, trade in silk increased from 9.8 million yen in the 1870s to almost 42 million yen in the 1890s.[2] When this trade commenced, Japanese silks were sent to Shanghai for re-export. At that time, Shanghai was a leading silk emporium in East Asia (a rival in this to Canton), and the Western merchants based there had developed some expertise in the trade. However, by the late 1860s both the P&O Line and the French *Compagnie des Messageries Impériales* had begun regular services to the principal Japanese treaty ports. Thus the Japanese export trade in silk was reoriented to flow directly either to Marseille, and thence to Lyon, which was the principal French market for raw silk and silkworm eggs; or to London, which was the centre for the world silk trade until the mid-1870s, when it gave way to Lyon. Silk weaving had a long history in England, and two localities stand out for their domination of this craft. The first was Spitalfield, a district just outside the City of London[3] that after 1685 had attracted a large population of Protestant refugees from France. These people, the Huguenots, established there a hand weaving trade of considerable importance, although it had gone into decline by the 1860s as a result of the industrialization of the weaving process and the importation of cheap French-made silk fabrics. After this the British industry shifted to the north, where it centred on Macclesfield in Cheshire. Here a factory-based industry developed that for a time would be the largest producer of factory-made finished silks in the world.

The American domestic industry would come to be concentrated in the northeast, particularly around Paterson, New Jersey, and neighbouring towns in Pennsylvania such as Allentown, Quakertown, and Reading. In these locales it focused initially on the manufacture of silk ribbons. Later, as power looms were introduced, the industry shifted to the manufacture of silk cloth.[4] Tariffs protected the nascent American silk industry until late in the nineteenth century, and this spurred the growth of domestic manufacturing. From the outset the American industry relied on Chinese raw silk, even though this product, because of its imperfections, was not well-suited to the requirements of the ribbon industry. As the industry gradually shifted to a higher level of technology and expanded to include the making of silk cloth, Japanese raw silk competed with Chinese silk as the source of supply, replacing it by the mid-1880s. In the 1870s, Chinese and Japanese raw silk began finding its way to the United States not via London but via San Francisco and Vancouver, from which points it could be shipped east by railway.[5] These developments made the American market an increasingly important one for the Japanese, whose producers began adapting their methods and improving the quality of their goods in order to meet American expectations rather than those of Britain and Europe.

It is important to contrast the Japanese industry with that of Europe, for the differences tell us much about the trade itself besides illuminating the changes that were beginning to impose themselves on the Japanese industry as a result of trading contacts and conditions. First, sericulture was well developed in Japan by the time the country opened to the West in the 1850s, so it was comparatively easy for that country to mobilize a supply of raw silk for export. Initially, Japanese silk was viewed very favourably in the London market because of its excellent quality and because it was cheaper than silk from China and Bengal.[6] However, in the rush to supply this new market, the Japanese ignored quality and consistency in thread size and in packing, and as a result, the reputation of Japanese silk had dropped significantly in the international market by 1868. European and American silk merchants, concerned about this deterioration in quality, lobbied the producers to arrest this decline. In 1868 the Yokohama Chamber of Commerce complained about "false trade labels, bad winding, irregularity of size and thread, dirtiness, lack of strength and the weight and irregularity of paper ties" and warned producers and shippers to correct their ways because the disruption of the domestic supply in Europe would not last forever.[7]

These developments pre-dated the opening of the treaty port at Hiōgo. Japanese raw silk would continue to be regarded as of lesser grade for several years after that event. Eventually, though, Japanese silks came to be regarded as superior in quality to Chinese raw silk in the export market; they were particularly appreciated for their very white colour, a characteristic that made them desirable for the production of light-coloured garments. As a consequence of this, and in the context of the crisis in the French and Italian industries during the 1870s, Japanese raw silk filled a particular role in silk weaving – namely, as weft rather than warp in the production of cloth.[8] Notwithstanding shippers' concerns about quality, some superior-quality silk was exported in the early years of the trade, and some districts in Japan were able to establish a reputation that made their product sought after in international markets.

The European and Japanese industries were at different stages of technological development. In Japanese raw silk production, thread was made by hand reeling; by contrast, the Europeans used mechanical power to rotate spools and to twist the filaments to produce thread. The latter technology permitted the production of raw silk of a more uniform standard in factories. Silk production in Japan, by contrast, was a traditional craft that was loosely organized across rural areas in which this was a local specialty. Finally, producers of Japanese silk tended to be found inland, isolated from foreign merchants, who were restricted from moving beyond the immediate environs of the treaty ports. This meant that merchants had to procure silks from Japanese dealers rather than from the producers themselves.

The new Meiji government, desperate to develop an export staple to balance the early dominance of imports, combined with some domestic producers who were concerned about the long-term health of the industry to encourage the development of Western-style filature factories using imported Italian machinery.[9] The promoters of these reforms resorted to competitions, fairs, and exhibitions to induce quality improvements among domestic producers. In addition, the government sent delegations abroad to study the latest techniques in silk manufacture. A Silk Inspection House was established in Yokohama under the control of the export merchants there; this initiative failed, however, owing to producers' resistance, and was abandoned in 1877 in favour of other attempts to bring regulation and quality control to the industry.[10]

Accordingly, manufacturing plants were established in Gunma, Fukushima, and Nagano prefectures to the northeast of Yokohama in the early 1870s, under the guidance of European engineers.[11] These tended to be locations where domestic production was well established and where entrepreneurial energy rose to the surface. Maebashi silks, which had their own characteristics and quality, were developed as a result of these strategies. However, insufficient capital forced many producers to resort to home-grown innovations when implementing

European technologies. Waterwheel power was typically employed, and wooden machines were fabricated in place of iron ones. Generally, these efforts suffered from a lack of managerial expertise as well as an irregular and insufficient supply of cocoons. These supply problems meant that factories often worked seasonally, making mass production difficult to achieve.[12]

Ironically, this supply problem was the unintended consequence of the rush to export high-quality silkworm eggs and cocoons to rebuild the industries in France and Italy; in that rush, not enough was done to increase the supply of these essentials to the domestic industry. Inevitably, then, some of these early enterprises went bankrupt before the industry was able to stabilize and consolidate itself. Throughout the 1880s, the industry sought to improve its cost controls and market evaluation, as well as its production methods, which were complex. Improvements in cocoon culture and the intensive use of young female labour, recruited from the nearby poorer agricultural districts, became increasingly critical to the industry. However, these changes took time and would not be achieved until near the end of the nineteenth century.[13]

Given the geographical location of Japanese silk production and the timing of its initial export trade, it is not surprising that Yokohama emerged as the centre for that trade. Several of the British, European, and American merchant firms that located there specialized in silk, applying the knowledge they had acquired in both Europe and China. But efforts to tap into the established Japanese domestic supply systems faced several hurdles. Sugiyama provides an incisive account of the procuring and trading of silk. Raw silk for export was entrusted by its producers to Japanese export merchants at Yokohama in exchange for bills of credit through local banks. Once those merchants received the silk at their godowns, they attempted to sell it at an optimal time and price by negotiating with Western merchants on the basis of samples. The price having been agreed upon, the Western merchants took the raw silk to their godowns to inspect it for quality and weight, without remitting any payment or issuing any documents. After the Japanese export merchant was paid, he then could settle payment with the producer, subject to deductions for interest, commissions, and other expenses.

For the Western firms, knowledge of the tricks of the trade was crucial. For example, Japanese export merchants apparently made it a practice to store silk products in damp godowns in order to increase the weight of the silk parcels. Thus experienced foreign merchants, before accepting delivery, made it their own practice to spread out the merchandise to dry before weighing and paying for it.[14] Throughout the period described in this book, foreigners and their consuls complained about the lack of commercial "morality" among the Japanese; by this they meant that the Japanese did not understand Western contractual principals – for example, the need to pay debts in a timely manner. In 1890, the British Consul at Kōbe complained that the merchants had condoned this for too long and expressed the wish that they would get together and demand "safe, desirable, and wholesome commercial habits" from the Japanese.[15]

For their part, Japanese export merchants complained that Western merchants frequently held silk in their godowns for several days before inspecting it, in the hope of more favourable market quotations, and that they often cancelled transactions, claiming deficient quality, if the markets remained dull. These practices left the Japanese export merchant to pay the costs of recovering the consignment. The Japanese also complained that American merchants were using the telegraph to fraudulently manipulate market price quotations in both America and Yokohama.[16] The business of trading silk was a perilous one for both producers and foreign merchants, given the multitude of markets and the international competition among producing regions. Wide fluctuations in silk prices and currency valuations meant that foreign merchants did not always profit from the trade. Moreover, the early 1880s were years of growing economic nationalism in Japan. In an effort to exert more control over the marketing of Japanese commodities, the Meiji government had begun to encourage their direct export; by bypassing the foreign merchants, the government hoped to wrest greater control of the terms of international trade.[17] We turn now to look specifically at efforts to initiate the silk trade at Kōbe.

Kōbe and the Export of Silk

The Japanese believed that foreigners held the upper hand in the silk trade. The reality, however, was more complicated: firms, even those with great expertise in the trade, struggled to establish profitability in the ports. One such firm was Aspinall, Cornes & Co., one of the first trading houses to establish itself in Kōbe, in 1868.[18] This firm had been formed in 1861 as a partnership between W.G. Aspinall, an expert in tea, and Frederick Cornes, an expert in silk. It was therefore a perfect marriage of the skills required to exploit the opening of Japanese trade. Cornes was born in Macclesfield, England, in 1837, the son of a silk mill owner of that city. As a young man, having reckoned that the prospects for silk in England were poor, he went to Shanghai in 1857 to work in the firm of Holiday and Wise & Co. There he remained until 1861, when he linked with Aspinall. The firm thereby formed continued until 1873, when it was dissolved upon Aspinall's retirement. At this point, Cornes was joined by other partners in the firm of Cornes & Co. Remarkably, the firm continues to this day, although its focus has shifted to other trading pursuits.

Aspinall, Cornes & Co. and similar firms expected that when Hiōgo opened to trade, it too would provide opportunities in silk. One reason for this expectation was that the Kyōto area had long been a centre of excellence in silk weaving and design of the very highest order, stimulated in part by the presence of the Imperial Court and its attendant religious and ceremonial life, for which silk garments were essential. British consular officials at Hiōgo noted, however, that the principal silk-producing districts associated with Kyōto lay to the east of Lake Biwa and that surplus silk destined for export was already moving overland from there to Yokohama. Similarly, woven fabric was being distributed to domestic markets within Japan by Kyōto manufacturers through existing networks and modes of conveyance. It was also noted that Osaka, the leading textile manufacturing centre, did not have a significant silk weaving industry, although its shops, as in Yedo, carried large inventories of silk and silk garments.[19] With respect to the domestic silk industry, the consul reported:

> The silks of those parts [Kyōto] are much prized by the Japanese themselves, and are woven by them at Kioto [sic] into the somewhat expensive Habutai stuffs which are worn in the palace of the Mikado, and are frequently used in making presents. The Hama crepe and the Omeshi crepe are also much esteemed, and are woven from the silks of the neighbouring districts. With a new demand the supply will probably increase. While accompanying Sir Harry Parkes on his tour of the west coast, I had an opportunity of observing the limited extent to which mulberry was cultivated on each side of Lake Biwa, notwithstanding the superiority of the silk produced in those parts. With a new market opened so near their door, it is presumed that the peasantry will turn their attention more to the growth of the mulberry.[20]

Table 5.1. Movement of silk from the port of Hiōgo and Ōsaka, 1868

	To Japanese ports	To foreign ports
First six months of 1868	228 bales (of 80 catties silk)	69 bales
	29,198 cards (of silkworm eggs)	
Second six months	615 bales (of 80 catties silk)	492 bales
	71,757 cards (of silkworm eggs)	10,760 cards

Source: Hiōgo and Ōsaka Chamber of Commerce, Statistical Report, 1868.

The first year of trade from the port of Hiōgo was sparse at best, but as the figures for the first six months of exports show (see Table 5.1), bales of silk and cards of silkworm were among the products that were exported. Most of the silk moved to Yokohama, and relatively insignificant volumes of these items were exported directly overseas in this initial period, reinforcing the central role that Yokohama merchant firms were playing in this trade. Perhaps more than others, the trade in silk depended on a certain expertise – specifically, in grading the processed silk yarns and packing and handling the silkworms destined for domestic and foreign producers. The value of this expertise in preparing such delicate commodities for export cannot be overlooked, and since few men were available to fill this role even in larger firms, the logic of making Yokohama the centre of this export trade must have been obvious from an early date.

In reflecting on this trade and its potential in 1868–9, the Hiōgo and Ōsaka Chamber of Commerce noted

> that on no account can the silks that are brought to this market be taken in the same category as those which the Yokohama market is supplied ... The principal classes of silk that have been and probably will be brought to this market are from the districts of Aechesan, Maggerhama, Mashtah, Sodai, Tamba and Shida [and] prices for such silk range from 1,400 to 1,600 [*boos* per *picul*]. Toward the end of the last season there was, however, some quantity of superfine Oshu, Ida, Goshu silk held in stock at Kioto and Ōsaka for the use of the native manufacturers, and the owners of these silks being tempted by the high prices that were known to have been paid by foreigners for such qualities, brought them to this market for sale and thus the average price of purchases during the last ten months must be estimated at $611 per picul.[21]

The report goes on to discuss the international supply situation: as a result of the failure of the crop in Europe and the drop in exports from China, prices had become inflated. Silk exports from the port in 1869 and 1870 generated significant revenues. In 1869, exports of raw silk to foreign destinations amounted to $188,610; to Yokohama, $173,870.[22] The following year the trade exploded: gross exports of raw silk, cocoons, and silkworm eggs amounted to $2,335,750, with around 80 per cent of that figure involving goods being moved to Yokohama.[23] Robust efforts to develop the trade at Hiōgo in 1869 and 1870 gave hope that Japanese producers and merchants might be poised to channel silk to the port. However, the foreign merchants seem to have acted cautiously at this critical juncture, due in part to uncertainties arising from the outbreak of the Franco-Prussian War. What had promised to be a buoyant market in 1870, with high prices early in the season, was checked by foreign buyers at Hiōgo, although not it seems at Yokohama. As a result of this timidity, suppliers thereafter bypassed the Hiōgo market and returned to directing their product to Yokohama.[24]

Returns for trade from the port of Hiōgo and Ōsaka in the years that followed, even after the European domestic industry suffered further setbacks in 1876, indicate that little silk moved through the port. Returns for the period 1877 to 1879 indicate that the value of silk shipped was never more than about $16,000 annually; by 1879 it had declined to the point where only 204 *piculs* of waste silk were shipped to Europe, with 15 *piculs* sent coastwise to Yokohama, for an estimated value of trade of $13,000, or less than 0.2 per cent of the value of exports from the port in that year.[25] In the 1870s, then, Hiōgo largely disappeared from the silk trade after making a promising start. References to silk as an important trade commodity for Hiōgo largely ceased as the decade progressed, although the British Consul speculated that once the railway was completed from Kōbe to Kyōto and then beyond to Ōtsu, silk from that area might again be channelled through the port instead of being directed by producers and local merchants overland to Yokohama.[26] Yet the progress being made toward linking this rail line to one being pushed westward from Tōkyō made it unlikely that Yokohama's dominance as Japan's silk port would be challenged even if the actual distances were to Hiōgo's advantage. So these prospects never materialized, and throughout the 1880s, silk's importance as an export commodity at Hiōgo would remain minimal. Nevertheless, the Japanese authorities made efforts to expand sericulture into parts of western Japan.[27] In 1886 the British Consul reported that "raw silk never figured as an important item in the foreign trade of this port ...

However, [there has] been lately initiated a movement having its objective the establishment of a silk market here."²⁸ The *Hiōgo News* reported in 1888 that a Public Inspector of Silk had been appointed at the port and that the cost of the appointee's salary was to be divided among dealers, the silk producers' guild, and the prefectural government.²⁹

Yet this trade failed to materialize, and Kōbe would have to rely on other commodities to sustain itself as a centre for international trade. Indeed, by the early 1890s, the overall trade attributed to Hiōgo – or Kōbe, as it was officially known after 1889 – would begin to catch up with that of Yokohama. Observers stressed repeatedly that if it were not for Yokohama's reliance on the silk trade and periodic spikes in the flow of this commodity through Yokohama, Kōbe's trade volume would have surpassed that of its rival.³⁰ Clearly, Kōbe's ascendency as a port must have depended on something other than silk. In fact, it was tea.

Kōbe and the Export of Tea

Before examining Kōbe's role as a tea exporter, it is important to build a picture of the broader patterns of tea production and consumption. At the time that Kōbe opened, China dominated the international tea supply system. The most important export markets for tea were Britain, the United States, Canada, Australia, and Russia. British imports of tea amounted to an annual average of 121 million pounds in the period 1866 to 1875, and almost all of this came from China. Thereafter, British tea imports grew steadily; meanwhile, the importance of China as the supplier weakened with the large-scale development of the plantation tea industry in British India and Ceylon during the 1880s. By the end of the century the annual average imports of tea by the principal importing countries were as follows: Britain, 279 million pounds; Russia, 107 million pounds; the United States, 80 million pounds; and Canada and Australia, about 23 million pounds each.³¹

In analysing these patterns of consumption, it is important to recognize that tea preferences varied considerably among these importing countries. Tea has three distinct types: black, green, and oolong. Each is differentiated in part by the method of production and preparation. During the treaty port era, 85 per cent of the Japanese teas exported were of the green variety; unlike the black and oolong varieties, these green teas did not undergo a fermenting process prior to export. Japanese tea had come to be preferred in the North American market by 1898; the black Chinese teas, and after 1880 the Indian and Ceylon teas, all of which were fully or partially fermented, were preferred in the British market.³²

When trade began in Japan, tea figured prominently in exports from both Yokohama and Nagasaki, but by 1865 Yokohama had established a clear dominance in the trade, a position it would retain throughout the treaty port era. As noted, American and Canadian consumers preferred green teas, and if Japan were to seize a share of these markets it would be necessary for merchants to dislodge Chinese green tea from them. It was recognized as early as 1865 that Japanese tea more closely matched American tastes. Also, consumption appears to have increased following the Civil War, due in part to the abolition of an import duty of 25 cents per pound on tea in 1872. Sugiyama notes that this lowered the price of tea relative to other non-alcoholic beverages such as coffee, consumption of which was substantially greater than that of tea in the United States at the time. In the last quarter of the century, market prices improved further as shipping costs were reduced. Other factors were also at work, one of which was that tea was favoured by settlers on the Great Plains, where the drinking water had a high calcium content. Japanese tea apparently masked the alkaline taste.³³

These economic and taste factors were only part of the story. Fashion seems also to have favoured the adoption of Japanese tea in North America, where it became the vogue as an "exotic" consumer good among immigrants who were eager to display their material and social advancement by acquiring foreign luxuries. These instincts were not manifest in Europe in anything like the same degree, but it was also the case that British merchants and investors aggressively sought to deflect consumer tastes toward the black teas because of the considerable stake they had made in the South Asian tea industry.

Tea exports from Kōbe grew remarkably in the port's first dozen years. As Table 5.2 shows, exports increased more than tenfold during this period, and as we shall see, tea became the major trade staple of the port, greatly outdistancing other commodities. In the process, the Foreign Concession at Kōbe became the principal site for procuring, collecting, and processing export tea, and many of its merchants developed premises devoted specifically to tea firing, grading, and packing.

One trading firm engaged in this business was Augustine Heard & Co., an American enterprise active mainly in Hong Kong and Shanghai. Heard & Co. was one of the first trading houses to arrive in Hiōgo when, in July 1868, it acquired a "beach lot in the Native Town fronting on the

Table 5.2. Tea exports from the foreign port of Hiōgo, 1869–83

1869–70	1,448,864 lbs of tea
1870–71	2,621,456
1871–72	4,129,170
1872–73	4,611,774
1873–74	4,562,525
1874–75	5,873,214
1875–76	6,082,036
1876–77	6,989,223
1877–78	8,789,627
1878–79	9,723,186
1879–80	13,710,439
1880–81	15,426,662
1881–82	13,048,185
1882–83	13,317,171

Source: Reports of the U.S. Consul at Ōsaka and Hiōgo, October 1884.

public thoroughfare" to ensure that it could commence operations at the earliest possible moment. Company records indicate that its investment in these premises was substantial. The acquisition of a dwelling for the agency cost $8,636; the erection of a stone godown, completed in 1870, another $4,456; a godown for tea firing and packing, $1,911; and a stone wall and fencing, $4,445. In all, the firm invested $15,450 in buildings to conduct a business that had yet to prove viable. Add to this a further expenditure of $2,228 to purchase lot #7 – one of the most prestigious Bund lots in the Concession – no doubt in expectation that the settlement would emerge as the principal location for foreign trading firms. This lot seems to have been acquired through an arrangement by which the firm advanced the money to an independent agent, Franklin Blake, who then established his own sub-firm under the auspices of Heard & Co. So as not to miss other potential opportunities in the Japan trade, and buoyed by the rising optimism that followed the depressed markets of the preceding two years, the firm in January 1870 paid $1,324 for a lot in Ōsaka's Foreign Concession.[34]

Unfortunately, the company's enthusiastic investment was not rewarded. The race to set up shop in the Native Town and to invest in buildings on leased land was soon seen to have been a mistake; foreign trade would clearly be centred in the Foreign Concession. Shortly thereafter, Heard sent its own agent, H.G. Bridges, to Kōbe to replace Blake.

Bridges soon realized the folly of the decision to locate in the Native Town, and reported that foreign traders had been obliged to conduct their affairs at the customs house in the Foreign Concession rather than at the "Western Customs House" in the Native Town. Moreover, the absence of a jetty near its premises in the Native Town had become a significant disadvantage for the firm, which now sought anxiously to cut its losses by selling its improvements to the property in the Native Town. As the firm's desperation increased, especially in the face of a slumping tea market, so did its willingness to sell these assets at a considerable discount. In February 1873 the firm advised Bridges to sell the property at anything over $3,000.[35] To this Bridges responded:

> We will be fortunate to get $3,000. I know this is out of proportion to rent formerly obtained from a Chinese tenant, but believe it will be very difficult to again rent it to a Chinese merchant as they have come nearer the concession and the situation is an awkward one for business being so far from a jetty and coolie hire is expensive here. To rent to Japanese should estimate the figure all in at $20-25 per month and not easy to find a tenant. The lot contains 3,700 sq ft and I suppose the buildings on it are worth $1,000 to $1,500. It is between 200 and 300 yards of the Railway station and may one day become valuable on that account. The Railway will probably be open to Ōsaka early next year.
>
> PS the property of Hudson Macomb near Mr. Blake's old place [in the Native Town], formerly rented to Chinese at $50 per month has been vacant two years and their agent now accepts $7 per month just to have it looked after. Believe our property is twice as valuable.[36]

Faced with the difficult market and impending losses on the sale of the Native Town location, Heard turned its attention to how it might shift its operations to the Concession lot that Blake had purchased on its behalf. Bridges argued that while the firm's Concession lot could be used for firing tea, this was impractical, since firing required close and constant attention and the property was 270 yards from the firm's present location. In a series of observations, he speculated about the cost of consolidating operations around the Concession location and the prospects for acquiring an additional lot adjacent to Blake's for this purpose. While lot #16 was available immediately to the rear of the present lot, the owners having returned to England, the price was likely to be high owing to the owner's "exalted idea of the value of Kōbe land."[37]

If finding the right location and establishing the appropriate mix of tea firing and storage facilities was a problem, so too was finding and maintaining staff. The records of Heard & Co. for the spring of 1873 reveal something of this problem as well as how firms trained inexperienced young men.[38] Bridges referred to the firm's junior clerk, John Gillingham, whom Blake had hired but who had not been informed that he was in fact joining Heard & Co. and consequently might be moved to other agencies within the firm's wider orbit. At issue was whether Gillingham might be relocated, since the Kōbe agency had also been sent another junior staffer, named Livingston. Bridges stated that Gillingham was a good employee, "not at all cheeky," who had bought a house lot and furniture and who was about to be married when the prospect of relocation was raised. Bridges noted that the need for staff varied seasonally, for the full attention of one man was required throughout the tea firing season, but thereafter the need was uncertain. He argued that it might be possible for the firm to get away with only one junior person at Kōbe if it abandoned some other business activities, such as serving as agents for Holt steamships. In other correspondence, Bridges made it clear that many of these young men were ill-prepared for the work when they arrived. Regarding whether Gillingham was essential, Bridges reported that only recently had Livingston done any work worth mentioning and that when he arrived, his services were not needed so he did nothing. When finally put to work, he had done well, but he still had much to learn. Bridges implied that the variety of work the agency conducted presented inexperienced men with a steep learning curve, albeit one that ensured they gained wide experience such that over time they became assets to the firm. This exchange concluded with Bridges declaring that he wanted to return home and that he believed the work could be handled by Gillingham and Livingston under the supervision of Cunningham, who had been Bridges's deputy and who was an old hand with Heard & Co., having earlier served in China at Hankow.[39] Nevertheless, it appears that by the fall of 1873, Gillingham had been released and had found work elsewhere in Kōbe. This meant that Heard & Co. would have to replace him with someone with experience in tea. The replacement was a man named Low, who would serve with Livingston as the agency's junior staff.[40]

We next examine Japanese tea production and processing in this era. The following is based on a report provided to the U.S. Secretary of State by the American Consul.[41] Tea was grown throughout Japan but chiefly in Suruga and Totomi, two districts surrounding the city of Shizuoka on the Pacific coast between Yokohama and Nagoya. Totomi was close to Yokohama, and much of what was produced there undoubtedly found its way to merchants in that port. There were, however, other localities closer to Kōbe, including the towns of Uji and Yamashiro, west of Ōsaka, and Goto, near Kyōto, from which Kōbe merchants might reasonably draw supplies of tea for export (see Map 0.2). In addition, there were tea-growing locations farther afield, such as the town of Tosa in Shikoku to the west, and Kaga, near Fukui; both were within 240 kilometres of Kōbe but also fell within the orbit of Yokohama.

Two varieties of tea, long- and short-leafed, were grown. Tea plants have a bush form, and farmers typically conditioned them to reach a height of three to four feet to maximize convenience for pickers during harvest. Plants typically matured at three years and were at their best from the fifth to the twelfth year, after which they required more care and manure than a younger bush, although it was believed that quality did not deteriorate with age. The plants were started from seed and required well-drained soils. For this reason they were usually grown on hillsides, where the land also happened to be cheaper. Growers terraced the hillsides slightly to prevent torrents of water on rainy days. It is a characteristic of the tea plant that some shoots bear smaller, harder leaves than others, and for farmers, this created a problem of grading during picking. However, Japanese growers were spared the problem of insects – something that challenged Indian growers. The first tea crop was harvested during a twenty- to thirty-day period beginning around the first day of May. The second crop was picked in June and July, and in some situations a third crop might be harvested. The picking was nearly always done by women, who typically harvested three or four pounds of tea per day.

Japanese tea, once harvested, had to be processed before it could be sold. The firing was carried out in stages.[42] The first of these, usually conducted by the growers themselves, involved steaming the leaves as soon as possible after picking by placing them in a wood-and-wire tray over boiling water for about thirty seconds. After this, the leaf was placed on a wooden table for a few minutes, then taken to the firing room. There, a bundle of leaves weighing about six and a half pounds was placed on a paper tray and fired in a charcoal oven until the weight was reduced to one and a half pounds. After this, men with bamboo trays shook the leaves, sorted them by size, then sifted and further sorted the leaves before spreading them on tables, where girls picked out seeds, stalks, and other rubbish. The refined tea was then sent to the nearest treaty port for sale.

Initially, teas for export were merely refired at the port; this was necessary in order to cure the leaf sufficiently for it to endure transport through the tropics and retain its essential qualities while in storage. This process alone required establishments with extensive facilities, as well as the financial capacity to purchase the tea and to pay for the labour and fuel required during the second phase of firing. Shippers soon grasped, however, that the leaf was improved by this effort. Thus after being purchased from the grower, the tea underwent a second and more carefully executed firing. There were two types of firing. In basket firing, the tea was placed on a tray in a bamboo basket and fired. The tea was stirred only occasionally, and the tray had to be removed at intervals to prevent scorching. This process took forty to sixty minutes. Pan firing, by contrast, employed pans heated by charcoal fires in a godown, with women conveying baskets of mixed tea to the pans, stirring the contents for thirty to forty minutes, then removing them and placing them in cooling pans for ten to twenty minutes, or in some cases taking them straight to the warehouse. Later the tea was sifted, packed, weighed, boxed and rattaned, marked, and shipped.

Clearly, firing was very labour-intensive work, whether on the farm or in the tea-firing godowns of the port. Women employed in the godowns worked from four a.m. to six p.m. for twelve to fourteen cents a day. Perhaps because of the seasonality and uncertainty of deliveries from the countryside, the merchant firms preferred to hire workers by either the day or the hour, on a first-come first-hired basis. Would-be workers therefore gathered at the godowns between two a.m. and four a.m. in hopes of being hired for the day. The presence of gangs of men and women seeking work came to be regarded as a nuisance by those living in the settlement because of the racket they created during their daily wait for employment. This was not the only tea-related irritant to the sensibilities of those residing in the settlement. The tea godowns themselves were generally considered to be eyesores: low, single-storied, soot-covered wooden structures. In later years one American writer would describe the scene inside the godowns: "half naked perspiring men and women bending over the great kettles and allowing the rain of sweat invoked by the high temperatures to fall therein upon the tea."[43]

A controversial aspect of the processing and marketing of Japanese tea was the application of artificial colour to the product, a practice that became common in the mid-1870s. Around 1870, consumers began demanding a higher colour than any natural process would furnish, and after some resistance, artificial colouring became the rule. Actually, there were two impulses at work here: some merchants were attempting to pass their product off as black tea; at the same time, the European and North American markets had apparently developed a preference for certain colours in tea, and tea firing firms in Kōbe accommodated them by colouring it. For example, Japanese teas, naturally blackish-green, were made to resemble Chinese bluish-grey "green teas" by the introduction of indigo and gypsum. Initially, merchants aimed to sell these adulterated teas in China; from there, they could be passed on to markets elsewhere as Chinese tea. In 1877 the American Consul at Hiōgo reported that 591 chests containing 53,500 pounds of tea had been shipped to Shanghai for re-export from China as domestic black tea. No doubt some of these adulterated products found their way to European and American markets. Some Shanghai merchants were reportedly involved in this activity with their counterparts in Kōbe, supplying Chinese labour to carry out these processes in Kōbe. It was also the case that coarse leaves could more easily pass undetected through the refinement and selection processes when coloured; this was another strategy employed by the tea merchants.[44]

The Japanese merchants saw this process as indicative of an unsophisticated consumer. Evidently they believed that anyone silly enough to want coloured tea would not care about the quality of the leaf. Lack of quality control in tea grading thus became a problem, and from time to time, overseas tea dealers complained about the poor product they were receiving. Unfortunately, the great distances and time lags between shipping and receipt left them little recourse but to try to improve standards from a distance by forwarding their complaints to the agencies in the port. During the first several years of the trade, these practices did not seriously affect the growth in exports; but in 1882, the matter was brought into sharper focus when the market finally reacted and trade slumped from its 1881 peak. In a report to his superiors in 1882, Stahel, the American Consul, indicated that these practices had finally caught up to the industry:

> This trade has gone from bad to worse until it has now become unsatisfactory both to the Japanese producers and the foreign exporters. Whether as a result of oversupply or of such deterioration in the quality of the teas shipped as tends to check consumption, the prices to which tea has fallen in

America are ruinously low, and if some improvement be not effected, this important commerce will be shunned by all who have anything to lose.

The Japanese government, recognizing the gravity of the situation, is urging producers in the country to reform their methods of preparing the leaf so as to furnish a better article for export and thereby reduce the excessive supply. One argument employed toward this end is that the law lately enacted by Congress against adulterated teas will, if strictly executed, exclude much of the inferior stuff.[45]

He went on to highlight the individual efforts being made to effect change within Japan by noting the remarkable articles on the subject that had recently been published in local newspapers. He urged that these steps be publicized in the United States so that buyers might become more discriminating. In predicting that a larger amount of uncoloured tea would be shipped in 1882, Stahel felt compelled to warn that "this movement will fail unless tea-drinkers in America can somehow be awakened to the fact that bluish gray and broken leaf is not the natural and proper form of the precious commodity."[46] He also stressed that pressure for uncoloured tea would result in greater discrimination in the picking and preparation of the leaf in Japan, thus affording consumers better tea at lower prices, and would revive the favour that Japanese teas had once enjoyed in the American market relative to the highly coloured teas of China. Demand for uncoloured "basket-fired" teas did increase in the United States as a result of efforts by the U.S. government to expose the adulteration practices that plagued the industry and to enact trade legislation aimed at preventing the entry of adulterated tea to the United States. There were efforts, too, to curb the practice of tea adulteration at its source; the *Hiōgo News* reported in 1888 that "nine more [Japanese] tea dealers have been fined" by the Japanese authorities for conspiring to sell adulterated tea.[47]

Tea firing was conducted in the Native Town as well as the Concession, but it was the foreign merchants, Western and ultimately also Chinese, who did the exporting. Although North America was the core market for these teas, British firms came to dominate the tea export trade. The record of exports from Hiōgo for 1880 indicates that eleven trading firms shipped tea. Of these, all but four were British. However, the three largest shippers were Mourilyan, Heiman & Co. (British); Paul Heinemann & Co. (German); and Smith, Baker & Co. (American). These three between them handled more than half the exports from the port (see Table 5.3).

Table 5.3. Shipments of tea by firm from Kōbe, 1879 and 1880

			1879	1880
Mourilyan, Heiman & Co.	British	piculs	25,266	24,404
P. Heinemann & Co.	German		20,277	21,932
Smith, Baker & Co.	US		19,659	15,511
Hunt Hellyer & Co.	British		8,686	13,900
Delacamp, McGregor & Co.	British		6,265	9,159
Browne & Co.	British		3,070	8,991
Cornes & Co.	British		13,308	8,796
John Gillingham & Co.	British		4,186	5,486
Mollison, Fraser & Co.	British		5,248	4,501
Walsh Hall & Co.	US		n/d	2,796
Fearon, Low & Co.	British		2,151	1,981
		TOTAL	108,117	117,457

Source: US Consular Reports, 1879, 1880.

The strength of British involvement in the trade reflects that country's overall dominance in the port as well as the fact that traders exploited mercantile opportunities wherever they could be identified. But there is also evidence that these tea exports were shipped in British and German ships to the American market, suggesting in turn that traders may have favoured their national fleets for the transport of goods.[48]

The merchants' capacity to engage in this type of activity is best understood by examining the business relations that characterized their firms. Most of those firms were commission agencies in that they pooled capital, entrepreneurial ability, and business connections with those of merchants in the home market; then, from their bases in Asia, they secured export goods and arranged their transportation to Western markets. Such firms succeeded or failed based on three factors: their ability to raise capital; their ability to apply specialized knowledge in identifying and procuring the goods they were exporting; and their ability to conduct this business with maximum efficiency and minimum risk. If the enterprise succeeded, the profits consisted of the commissions paid to the firm, with what remained going to the principals who had provided the capital. Thus market intelligence and the ability to recruit and hold on to investors were essential to a firm's survival. It is arguably the case that the Asian trade of the 1840s and 1850s, when opium was the principal commodity, had been a bonanza for investors. During the 1870s and 1880s the Asian trade settled into more reputable and controlled patterns, with smaller margins. But it was also the case that domestic investment opportunities in the United States increasingly induced American investors to focus their capital at home. This left the British and in some sense the German merchant houses to dominate the ground in Kōbe and Yokohama.

As in any agricultural commodity market, fluctuations were created by the vagaries of weather as well as by producers' reactions to perceived market conditions. There is some evidence that prices in the American market were also affected by falling freight rates toward the end of the 1870s. One factor that contributed to the decline in freight rates was the growth of the "case oil" or kerosene trade, which provided shipping companies with a valuable outward-bound commodity for sale in Japan. This provided a way to balance the valuable homeward-bound tea cargoes, particularly for ports such as New York and Philadelphia on the east coast, which were close to the kerosene production centres.[49] Moreover, these goods increasingly were carried by steam-powered vessels, which used the Suez Canal as a matter of course. Thus by the beginning of the 1880s, products like tea were reaching markets more quickly and safely than ever before.

Yet the trade was still volatile. The American Consul reported in the autumn of 1880 that there had been a good profit to producers, who that year had sold an unusually large crop at good average prices. He advised, however, that the situation had been bad for shippers, for the high American prices early in the year had induced them to buy too eagerly, with the result that they had become overstocked. Under the stimulus of good prices, tea production in Japan had soared in 1880. The ongoing depreciation of the Japanese paper currency had also contributed to this result, given that sales to foreigners were always in silver. He noted that the market having subsequently fallen to an unusually low level, production in the following season would probably be less, or leaves prepared for export would have a smaller proportion of the coarser type.[50] The following year, Stahel, the American Consul, reported that the new tea season had opened 9 May and that the first purchases had been made for $28 to $35 Mexican per *picul*. Soon after buying began, the price advanced by $3.00 per *picul* even though the leaf was less satisfactory than the previous year, owing to weather. But the crop was expected to be large, he added.[51] As Table 5.2 shows, tea exports peaked in 1880 before settling at a slightly lower level in the following years.

As prices and production stabilized after the bonanza of 1880–1, producers and shippers became much more sensitive to issues of quality, grading, and packing. It seemed that quality was slowly becoming a consideration in the American market. It would be wrong, though, to interpret this as a natural evolution of market savvy on the part of shippers, for in 1883 the U.S. Congress had enacted legislation prohibiting the importation of low-quality tea, which severely cut the flow of ordinary teas from both China and Japan.[52] Curiously, shippers often retreated toward cheaper grades of teas when markets softened; it was the producers who led the movement to deliver "medium" grades as a means to improve their own market returns.[53] It seems that the Japanese had become more sophisticated in their efforts to influence market conditions abroad. But the Kōbe tea merchants soon embraced these strategies as well. In 1891, for example, the British Consul reported that while the tea crop was late due to spring weather, traders recognized that stocks overseas were low; thus he forecast a robust demand, with dealers seeking to get higher prices for the early contracts at a time of low exchange rates. The early crop was abundant and of good quality, and there was greater effort to pick the best leaf before it could degrade. This meant that the volume of tea was down but that prices were higher – something that heralded a new approach.[54] The consul also reported on a meeting of the Kansai Tea Dealers Association where the discussion focused on curtailing abuses in business practices and on the need to pay close attention to preparation and packing.[55] These developments perhaps reflected the fact that Japanese teas were facing growing competition in the North American market now that India and Ceylon were heavily promoting black teas. This campaign had begun with the presence of those competitors at the World's Columbian Exposition held in Chicago in 1893, after which they began to grow their share of the American market, so successfully in fact that Japanese tea shippers shifted their focus to the Canadian market.[56]

These changes can be tracked by noting the principal destinations for tea within North America. As Table 5.4 shows, between 1874 and 1884, New York was the dominant market, with Chicago and San Francisco in second and third place. Teas headed for Canada were landed principally at Montreal. A closer examination of the receipts of tea at these destinations shows that the overall fluctuations were not consistently felt at these points of entry. The New York market seems to have been the most volatile. Receipts of tea at New York increased by 17 per cent between 1879–80 and 1880–1, but fell by 36 per cent the following year. In comparison, receipts at San Francisco rose by 44 per cent and 43 per cent for the same years, suggesting perhaps that the Western regional

Table 5.4. Destination of tea exports from Hiōgo, 1 May–30 April 1884

Seasons	to NY (lbs)	to San Fr	Chicago	Canada	Total
1874–75	5,915,601	812,962	297,926	182,084	7,208,573
1875–76	6,367,884	384,392	323,743	16,200	7,092,219
1876–77	5,906,299	100,463	822,015	39,684	6,868,461
1977–78	7,102,911	150,653	1,315,537	120,526	8,689,627
1878–79	7,950,361	72,453	1,700,282	138,220	9,861,406
1879–80	8,647,372	58,359	2,407,299	2,597,409	13,710,439
1880–81	10,135,495	84,321	2,651,302	2,555,544	15,426,662
1881–82	6,527,414	121,026	2,312,700	4,087,045	13,048,185
1882–83	7,094,337	324,241	3,316,368	2,577,225	13,317,171
1883–84	6,258,887	619,868	2,817,748	4,253,549	13,950,052

Source: NARA, micro 460: 5, US. Consular Report, Patton to Secretary of State, 1884.

market remained buoyant regardless of price, quality, and other factors. The Chicago market, which probably also served the Western frontier, rose by 1 per cent before declining by about 13 per cent the following year.

In Canada the pattern was entirely different. Receipts declined between 1879–80 and 1880–1 by about 2 per cent but rose by a remarkable 60 per cent the following year. Indeed, the British Consul noted that Canadians were consuming more tea per capita than Americans.[57] Also, as much as one-third of the tea entering through American ports was destined for the Canadian market, so Table 5.4 may not provide an accurate picture of consumption patterns.[58] Canadian- and American-bound teas were increasingly moving in Canadian ships. The Canadian Pacific Railway was completed to Port Moody near Vancouver in 1886. This linked Pacific ports to eastern cities such as Toronto and Montreal, as well as to American cities. As soon as the line was completed, to show that their railway made economic sense, the CPR had chartered several sailing vessels carrying tea, silk, and curios from Japan. These vessels arrived at Port Moody within days of the first train reaching the Pacific coast.[59] By 1887 the British Consul was able to report that over 4,900,000 pounds of tea had been shipped by the CPR's own small fleet of steamers to Vancouver for eastern Canada and the United States. Moreover, 6,700,000 pounds had been shipped from Yokohama to San Francisco by steamer, and an additional 1,200,000 pounds had gone via sail to other ports on the west coast of North America, while an estimated 5,700,000 pounds had gone to America on British ships via Suez.[60]

Table 5.5. Destinations of tea shipped from the port of Hiōgo in 1894

USA	New York	6,205,838 lbs
	Chicago	5,990,324
	San Francisco	140,383
	Kansas	21,323
	Sioux City	19,225
	Detroit	19,200
	Milwaukee	9,044
	Cleveland	6,755
	St. Louis	4,822
Canada	(undefined)	2,289,278
	Hamilton	254,547
	Toronto	233,592
	Montreal	216,099
	Winnipeg	9,449
	Vancouver	982
Other	Vladivostok	44,587
	Bombay	1,333

Source: *Shogio Shimpo* (Commercial News) as reproduced in the *British Parliamentary Papers*, Japan Embassy and Consular Reports, IX:53.

As Table 5.5 indicates, teas shipped from Kōbe in 1894 had various destinations. Fully half the teas entering the United States were apparently headed for the West and Midwest. The picture is less clear for Canada, where a large "undefined" classification masks the ultimate destination. What is most striking, though, is the near irrelevance of markets outside North America: Kōbe's tea exports – indeed, all of its exports – were strongly tied to the North American continent. There is little doubt that from the moment Kōbe opened, tea was "king," and that this trade item had profoundly conditioned several aspects of the settlement. Land use, the economic nature of much of the built landscape, and the rhythms of economic and social life all reflected the centrality of this trade in the lives of the foreigners who resided there.

Chapter Six

The Morphology of the Settlement and the Development of a Pleasing Townscape

All treaty ports required infrastructure, including – crucially – port installations, but also including landward transportation systems for moving goods and people to and from their increasingly modernizing hinterlands. Who was responsible for these investments, and how did foreign communities shape these developments? In exploring these issues for Kōbe, it will also be important to track that settlement's form as it evolved, including developments immediately proximate to the Concession, such as in the Native Town and on "the Hill," where many in the foreign community lived and worked. Much of this exploration will focus on the architectural character of the settlement – that is, on whether the forms imposed on Kōbe reflected norms that were current in the various home countries of the residents, or whether instead they reflected an emerging hybridization, what might be thought of as a "Eurasian colonial" style. These observations will lead back to a consideration of conceptual models of the mercantile port city, with regard to whether Kōbe followed one or another of the patterns or models considered in chapter 1.

Development of Port Infrastructure

Improvements to Kōbe's port facilities were essential to trade there from the outset. As a trading hub at the eastern end of the Inland Sea, Kōbe exploited its advantage as a sheltered deepwater port and its ability to service Ōsaka and Kyōto and their respective economic hinterlands. In the minds of many, there must have been a nagging recognition that at some point the far stronger economic forces represented by the great industrial city of Ōsaka would take steps to overcome the hurdles to navigation that prevented that city from capturing the international trade that Kōbe had so successfully developed. In a port where success at trading rested on thin margins and where firms repeatedly unravelled and remade themselves, it was generally presumed that engineering projects of the sort needed to improve port infrastructure were the responsibility of the central government, in this case the Meiji regime. Western interests in Kōbe avoided making any large-scale strategic or entrepreneurial investments in transportation infrastructure, or in the building blocks of industrial production apart from a few comparatively small enterprises. These were, after all, mercantile traders, and most of their firms were small and sufficiently footloose that they might pick up and move to another location if better trading opportunities were perceived to exist elsewhere. At best, then, the foreigners' investments on the ground were largely limited to the comparatively small footprint of godown, tea firing factory, house, and office – the necessities of an import–export enterprise.

Once trade began, the Japanese authorities did make small improvements in the harbour and port infrastructure as an impetus to trade. In March 1870 the editor of the *Hiōgo News* reported that at the "camber" west of Kōbe in the Native Town, the Japanese authorities had made the dock square by filling in a corner. It is not clear which location the writer was referring to here. Perhaps it was the wharf that came to be known as the "Merikan Hatoba" at the foot of Nishi-machi, or what the foreigners popularly called Division Street, so named because it divided the Concession from the Native Town. It seems more likely, however, that he was referring to the site of the "Western" Customs House at the western end of the Native Town. If this is correct, their efforts probably mirrored steps they had taken earlier to create a safe landing place and small-boat harbour at the eastern end of the Concession to serve the "Eastern" Customs House, which serviced the foreign traders exclusively. The project did not receive a resounding endorsement from the editor, who reported: "Some godowns have been moved onto the street blocking it, a fence now obstructs a passage ... Many foreigners now own property [in this vicinity] and are living in the place."[1]

Because foreign trade was to be conducted through the Eastern Customs House, and because foreign interests always vigorously protected their own fragile property rights, the writer called upon the foreign community to resist the actions of the Japanese authorities in this matter. Later, in July 1870, the editor reported the launch of the first of four dredges imported from England by Messrs Alt & Co., a firm based in Ōsaka. These might well serve the needs of those anxious to improve Kōbe's port, but they could also be put to work dredging Ōsaka's harbour.[2] Concerns about improving Kōbe's port now took on greater urgency, and the resulting pressures would percolate steadily throughout the 1870s, albeit without appreciable results, no doubt because of the downswing in the port's general trading economy during that decade.

By the end of the decade, pressure was again mounting for a modern docking and cargo-handling facility suited to the port's increasingly important steam cargo and passenger vessel traffic. Such a facility would reduce the inefficiencies involved in using lighters to offload cargoes from these and other vessels; it would also address the future need for much more waterfront warehousing than had so far been built. It is significant that this project's champions were two Japanese entrepreneurs, Godai and Okoboshi, who organized a company, prepared plans, and submitted cost estimates. Thereafter, however, progress stalled until late in 1881, when there was optimism that their efforts would at last get under way.[3]

In the meantime, Mr Takahashi, Kōbe's Commissioner of Customs, sought to build a better wharf at the Eastern Customs House in order to assist the foreign trade that was centred there.[4] The proposed T-shaped wharf would be 120 *ken* (216 feet) in length. An application was made to the Japanese Home Department for permission to proceed; a foreign engineer employed by the Bureau of Public Works surveyed the location; and the prefectural government and Kōbe's customs officials all became involved in the initiative. Part of their plan involved laying down a rail spur to link the wharf to the Imperial Railway at Sannomiya, just above the upper end of the Concession.[5] It is not clear how Takahashi proposed to route this link, but it almost certainly would have required some encroachment on existing land within the Concession or on the Recreation Ground to the immediate east. There is no evidence of tensions within the foreign community over this proposal, but it does appear that this project, for whatever reason, was never realized.

Interest in port improvements now turned to Godai and Okaboshi's Kōbe Pier Co. project, which was centred on the Western Customs House, outside the Concession at the front of the Native Town. Plans for the No. 1 and No. 2 wharfs were put in motion, and negotiations to purchase the ground occupied by the foreigners' Kōbe Regatta and Athletic Club (KRAC)[6] were completed. Work commenced on this project at the end of 1883; it was completed in November 1884. The pier was 500 feet long and 42 feet wide, and it offered a depth of 23 feet at the outer end at the lowest spring tides and of 20 feet at a point 360 feet to landward. Six hundred feet from the pier's outer end, spaced 800 feet apart, were two buoys, each secured with two 3-ton anchors to which vessels could be moored. The pier was built of iron "screw piles," which were sunk more than 30 feet below the seabed. The piles were connected by wrought iron girders. Four fireproof godowns were constructed near the pier, each able to hold 1,200 tons of merchandise. There were also two large wooden structures for receiving, sorting, and delivering cargo. These buildings were connected to the pier by tramlines consisting of thirty cars, each of which was capable of carrying 5 tons. There were also assorted buildings for offices and customers. It was estimated that the working capacity of the pier would be 90 to 100 tons per hour. Charges were projected to be moderate: tea handling was set at 4 cents per chest, with special rates for large exporters. Other products were to be charged from 1.5 to 3 cents per *picul*.[7]

It appears that after this terminal was built, many merchants continued to land cargo by lighter, even when this consumed more time. This practice was no doubt entirely related to the requirement that foreign trade be conducted through the Eastern Customs House. Nevertheless, the superior berthing facilities for steamships provided by the Kōbe Pier Co. eventually found favour, particularly with the British and French mail steamers when these lines began using larger vessels in the late 1880s.[8] At the same time, additional storage sheds were being built near the Eastern Customs House in response to the growing volume of goods moving through the foreign port.[9] However, merchants in Kōbe must have seen this development as a mixed blessing, because of the distance between the new pier and the Eastern Customs House, where foreign trade was conducted. This preserved the settlement's visual harmony and land-use integrity, but it also resulted in inefficiencies because it required a great deal of unnecessary movement of goods back and forth between docking and customs facilities. All of this stemmed from the Japanese insistence that foreign traders restrict their activities to the Eastern Customs House.

The Kōbe Pier Co. improved the operational efficiency of the port and in so doing captured its mail steamer and passenger traffic. Apparently there was less concern about improving the infrastructure for the domestic trade through Hiōgo. The bulk of the existing domestic port lay to the west of the Kōbe Pier Co.'s facility, and it would be almost a decade before similar developments were undertaken there. These activities began in the summer of 1888, with the *Hiōgo News* reporting that one of the piers built by the Kōbe Pier Co. was being lengthened.[10] In September, merchants and mariners were informed that the extension had been completed and was ready to receive steamers. Meanwhile, a new pier would soon begin construction at the domestic port of Hiōgo at Benten no hama under the auspices of the Japanese Bureau of Public Works; however, this initiative soon became embroiled in controversy over the leasing of land and the entry of the Yokohama Engine Works to a portion of this site. These problems took some time to resolve; as late as 1892 the British Consul reported that although the domestic port had nominally been opened to foreign trade on 1 October of that year, few shippers were using it because it lacked the facilities associated with the foreign port. Nevertheless, it seems that the trade conducted by both domestic traders and foreigners had begun to converge on the wharves near the Western Customs House. This perhaps foreshadowed the blurring of the distinction between the activities and interests of these two communities of traders. As the treaty port era came to an end in 1899, port activities shifted easily to this western location, which provided an almost seamless linkage to the Imperial Railway through the main rail terminal for the City of Kōbe (see Map 6.1).

The Coming of the Railway

One of the bold triumphs of the early years of the Meiji regime was that it rapidly constructed railways. For Westerners, particularly traders, Japan's overland communications were utterly inadequate and frustrated their economic aspirations. The Japanese had long before constructed an impressive national trunk road that connected centres such as Tōkyō, Kyōto, Ōsaka, and Hiōgo, but there was no system of stagecoaches to expedite the movement of people or mail over this route. Even the land connections between Hiōgo and Ōsaka were viewed as poor, and an initial plan by Wm. Rangan & Co. of Yokohama to establish a coach service between the two centres in January 1868 was quickly abandoned because the roads out of Kōbe were so narrow.[11] Foreigners resorted to pony-and-trap communication, and a *jinrikisha* service was later introduced.[12] In 1870, a telegraph service was established between Ōsaka and Hiōgo.[13] As this new technology spread throughout Japan and as these connections allowed traders and others to link with the expanding international telegraphic service, the sense of time/distance

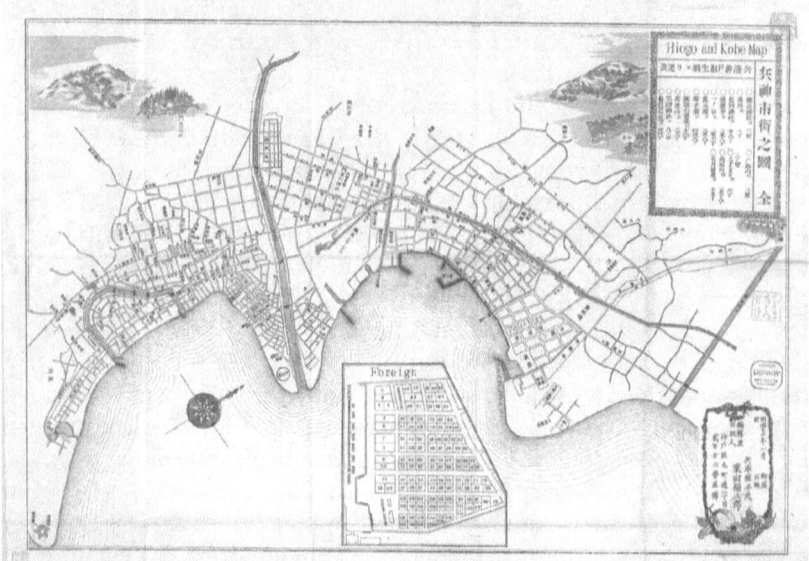

Map 6.1. Map of the Concession and surrounding areas, ca. 1880. Note the contrast in the density of streets in the Native Town, the railway route encircling the urban area, and the development of harbour facilities in Hiōgo; also noticeable is the semi-rural settlement on the Hill, beyond the railway.
Source: Pocket map by Awata, Fukusaburo titled Shinsen kaisei Heishin shigai no zu : zen, published 1880. Courtesy of the C.V. Starr East Asian Library, University of California at Berkeley.

isolation was significantly altered. Nevertheless, the most important improvement in internal communications was the railway.

Japan's fascination with railways was undoubtedly planted early, when Perry's second expeditionary voyage to Yedo in 1854 brought with it an operating quarter-scale model of a railway on a loop of rails approximately 100 metres in length. There is no question that railways were perhaps *the* icon of modernity in the middle decades of the nineteenth century, and the fact that Japan succeeded in building one (however limited) within five years of the Meiji Restoration was a clear indication that the country had embarked on a bold quest to catch up with the industrialized West. Not surprisingly, the initial routes connected the key population and economic centres that then, as now, formed Japan's heartland. The intent was to build the Imperial Railroad to connect Tōkyō with Yokohama, Nagoya, Kyōto, Ōsaka, and Hiōgo, before fanning it out into the hinterlands of these hubs. To accomplish this, the authorities applied a strategy that was widely employed for a number of other aspects of their national transformation: after assessing the technology and expertise of other nations, and determining which best served their aims, the Japanese authorities recruited foreign consultants to help implement their objectives. In the case of the railway, British experts were selected, and plans were set in motion accordingly.[14]

As with other national railway-building schemes, the authorities inevitably had to contend with land speculation along the projected routes. Kōbe was not immune from this, and some foreigners apparently assumed they would be able to cash in by selling their lots when the railway entered the Concession. Others purchased lots on the Hill, expecting to profit from selling them to the railway when the route was laid through this part of the settlement.[15] The foreign community's behaviour was undoubtedly a natural response to an opportunity, but given that profits from trade were slow to develop in the port before 1874, the desire to profit from investments in land and buildings would have been stronger for some. At another level, the railway would have significantly altered the nature of the Concession. At its deepest point, it extended only four blocks inland from the shoreline, which meant that a railway line across any part of it would seriously degrade and divide the community it served. In the surviving record there is surprisingly little discussion or opinion about this. The authorities responded to the foreigners' speculative activities by running the rail line along a route on the Hill different from the one the speculators had bet on, thus avoiding the high prices they were seeking.[16] The route selected led from the east across the top of the built-up area of Hiōgo, with a station at Sannomiya on the northeastern corner of the Native Town. In time, the railway would establish its main station in Kōbe farther to the west beyond the Concession at Benten no hama. Adjacent to this would be a large marshalling and locomotive service yard, as well as a railway pier for loading and unloading goods. Locating the line at Sannomiya, between 150 and 300 metres north of the Concession, ensured that a broad swathe of largely undeveloped land would serve as a buffer between the settlement and the railway.

The building of the Imperial Railroad required a significant infusion of foreign experts and personnel. A base of operations was established near the route, immediately northwest of the Concession. Work on the line was carried out in sections, the ultimate goal being to link these sections to a trunk line between Tōkyō and Ōsaka. At the end of July 1876 the American Consul, Nathan Newitter, reported that the extension of the Imperial Railroad between Ōsaka and Mukomachi, a town 6.5 kilometres from Kyōto, had been opened to the public, thereby completing a distance of 74 kilometres. This was said at the time to be the longest stretch of railway in the Far East.[17] Beyond Kyōto, en route to Ōtsu and Tōkyō, yet another leg had been officially opened by the Emperor himself, and by 5 September 1876 the intervening links had been connected so that rail travel was now possible between Kōbe and Ōtsu on Lake Biwa.[18] By October the American Consul reported that 1.3 million passengers had already been carried on this line, along with large volumes of freight, as well as troops.[19] The fact that this had been accomplished using British rather than American expertise was a particularly grating matter for the highly partisan Newitter. This rivalry was exacerbated when Newitter, as the ranking consul at Kōbe, was delegated to address the emperor on behalf of the Municipal Council on the occasion of the opening of the line at Kōbe.[20] The completion of the link between Ōtsu and Nagoya, and then on to Tōkyō and Yokohama, would prove a greater engineering challenge, for the line would have to cross more mountainous terrain, and this work was still in progress in 1888. Even so, feeder lines were proposed through Ōsaka to Nara and Wakayama. To the west of Hiōgo, the Sanyo rail line from Kōbe to Ōkayama was already being laid and would eventually link to Hiroshima.[21] By November 1888 the connection between the Sanyo line and the Imperial Railroad was eagerly anticipated.[22]

How important was the construction of the rail system to the trade conducted by foreigners at Kōbe? It is difficult to find contemporary comment on this. The rapid movement of people and information within the hinterland of the port must have been of some benefit. However, the procurement of tea and other trade goods was in the hands of Japanese traders, and it is not at all clear whether they encouraged farmers and other tea factors to bring their products to the port by rail. The Imperial Railroad line passed to the west of the primary areas of tea production, but tea could probably have been moved to the stations on this line with relative ease if the freight rates were competitive with other, more traditional modes of transportation. For the foreign merchants themselves, the railway probably facilitated personal short-term movement between Kōbe and Ōsaka. However, foreigners required internal passports in order to travel more than 40 kilometres beyond the Concession, so they apparently did not do so, by rail or by other means, except when absolutely necessary. Nevertheless, internal travel did take place; the *Hiōgo News* reported in July 1888 that eighty passports had been issued to foreigners for travel to Kyōto.

Relocation of the Ikuta River

One concern that confronted the foreign community soon after the Concession was opened in 1868 was the threat of flooding posed by the Ikuta River, which flowed close to the Concession's eastern boundary. Like all of the rivers flowing out of the Rokkō Mountains, the Ikutagawa had a short, steep gradient, and as a consequence, any heavy rainfall in the upper reaches of the watershed resulted in an extremely powerful run-off as the water sought an outlet to the sea. Japan is subject to occasional typhoons and to a pronounced spring monsoon of four or five weeks in May and June. All of the rivers in and around Kōbe generated potentially violent flood surges during these weather extremes. Over time, many of these rivers had created what fluvial geomorphologists call "raised beds." What happens is that repeated flooding causes the river to deposit its load of eroded material in the lower and flatter reaches of the river bottom. Moreover, when the flood water overflows the river's banks, silts are deposited along those banks. Repeated episodes of this deposition create levees – that is, raised banks that resemble dykes. Under normal flow conditions, the levees contain and channel the river. During much of the year, the river deposits its erosional bed load on the river bottom in the lower reaches, and over many years, this raises the river bottom. Over long periods, the natural processes of deposition and flood and consequent levee formation may cause the river's entire course to rise above the base level of the surrounding land. When this happens, the impact of floods on human settlement can be even more severe, since once the flood water spills over the levee, it has no means to drain back into the river as the flood pressure abates. If the river's erosional action causes a breech in the raised bank, flood water is likely to spill rapidly onto the surrounding flood plain, where it may cause considerable erosional and property damage, and leave the surrounding land under water until it can be drained by other means.

These conditions characterized the Ikuta River. During the first years of the settlement, it was recognized that the river's banks were in such poor condition that they frequently collapsed. Flood was a serious concern not only for the occupants of lots near the river, but also for those who were establishing activities outside the Concession to the east of the river. One of the most sensitive matters related to the foreign cemetery, which lay beside the Ikuta. The spectre of the river unleashing its erosional force and exposing the remains of those buried in the cemetery undoubtedly weighed heavily on the authorities, both foreign and Japanese. But there were other problems and consequences as well. Also subject to the ravages of the river's flooding were the International Hospital, the livery stables, the cattle yards and abattoir, and the growing assemblage of bakeries and other supply businesses that had taken root there since 1868. Another concern was that the extremely wide river bed would present an engineering challenge when it came to constructing a railway bridge that would link the region with Tōkyō.

In 1871 the Japanese authorities decided to relocate the Ikuta River about 400 metres farther to the east; this would be accomplished by digging a straight channel from near the river's headwaters at Nunobiki right down to the sea. This significant engineering feat was accomplished in a remarkably short three months.[23] The old river bed was then converted into a wide road, named Kano-cho after the Japanese contractor who had carried out the work.[24] Now that the river had been moved and stabilized, the tracts of land immediately to the east of the settlement could be used for economic and recreational activities that could not be accommodated within the Concession. The development of the Recreation Ground will be considered in greater detail in a later chapter.

Intra-Concession Relocation of Business Premises, 1868–88

It was characteristic of the ports that there was a lack of continuity of business firms both large and small. An analysis of occupancy data, albeit imperfect, reveals that only 20 of the 126 lots within the Concession were in the same hands in 1888 as they had been at the time of original sale.[25] Significantly, and not surprisingly, continuity was greatest for those lots at the front of the Concession along the Bund and the first two blocks of Kyo-machi – the sectors of the town that drew the most established firms and that generated the greatest investment in buildings. Continuity fell off sharply along the backstreets behind these two principal streets. However, a number of smaller trading firms established themselves in this less desirable sector, particularly at the rear of the Concession, where they erected a number of godowns for tea firing, some of which ranged over adjoining lots. Mixed in with these were offices, stores, and services, as well as public and institutional buildings such as the Municipal Hall, the Union Protestant Church, and the Roman Catholic Church and orphanage.

Piecing together the comings and goings of businesses in Kōbe reveals the transience of many of the enterprises there. A few, such as that of Mr Cabeldu – who seems to have pursued several lines of business including tailor and haberdasher, as well as other commercial activities – remained in place throughout the period. Other businesses moved from location to location within the Concession as they refined their business or passed it off to a successor. Monsieur Gandaubert, for example, established a hotel and restaurant on lot #15 in 1868; by 1872 this was known as the Hôtel des Colonies. By 1875, it had been relocated one block north, on Naniwa-machi on lot #56, and was being operated by a Monsieur Reymond in what appears to have been a sizeable three-storey brick building. The Hôtel des Colonies continued under this name at least until 1888. In the meantime the original hotel building was briefly taken over by another operator and renamed the Astor House; after that, it seems to have been transformed into a general store operated by Messrs Skipworth and Hammond. But this business also had relocated by 1888, to Cabeldu's considerable retail emporium on lot #30, a site that seems to have included the services of barbers and hairdressers.

Two other hotels bear noting: the Oriental and the Hiōgo, both of which were substantial and enduring operations. The Oriental had pride of location, on lot #80 of Kyo-machi adjacent to the imposing edifice of the Hong Kong and Shanghai Banking Corporation. Operated by the Frenchman L. Bégeux and his wife, the Oriental was evidently *the* place to stay for discerning travellers. Kipling heaped praises on this hotelier:

> Let me sing the praises of the excellent M. Bégeux, proprietor of the Oriental Hotel, upon whom be peace. His is a house where you can dine. He does not merely feed you. His coffee is the coffee of the beautiful France. For tea he gives you Peliti cakes (but better) and the *vin ordinaire* which is *compris*, is good. Excellent Monsieur and Madame Bégeux![26]

Madame Bégeux was among the first wave to settle in the Concession; she seems to have plied her trade as a confectioner until able – apparently with her husband – to establish the Oriental sometime before 1875.

122 Opening a Window to the West

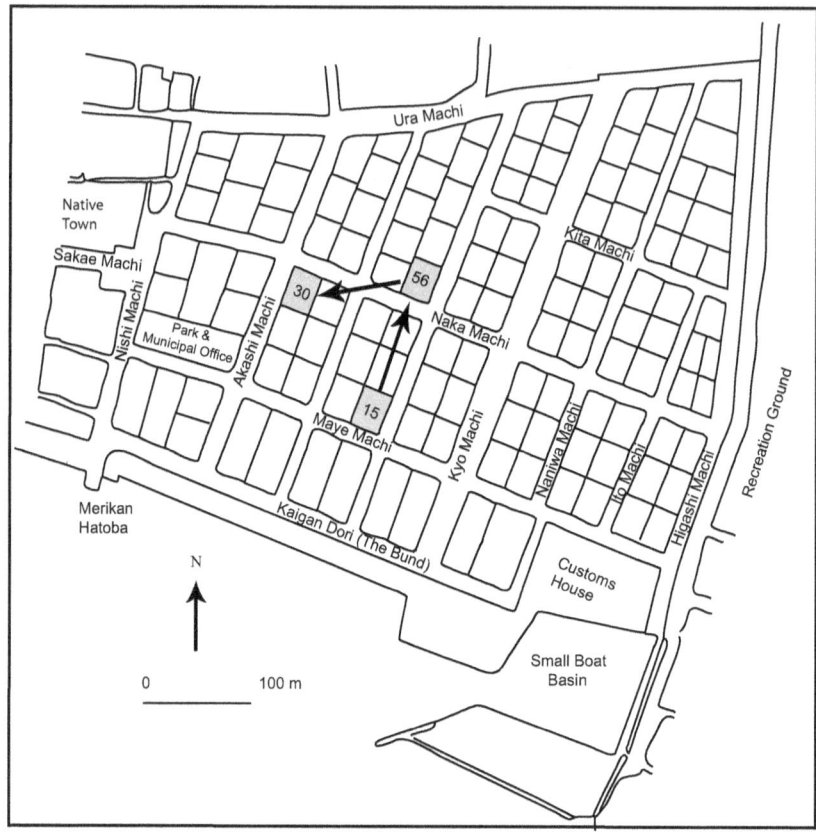

Map 6.2. Shifting locations of selected business, 1868–88. Note: Lot #15, ca. 1868, was the premises of Gandaubert's hotel; ca. 1872, those of the Astor House hotel; ca. 1875, those of the Skipworth and Hammond General Store. Lot #56 in 1875 was the premises of Hôtel des Colonies, operated by M. Reymond. Lot #30 in 1888 was the premises of Skipworth and Hammond General Store.
Source: Author.

The Hiōgo Hotel was founded in 1868, one of four hotels to be established as the port was being thrown open to traders. It was located first on the beach east of the Concession; a more substantial building was later constructed on the Native Bund, just outside the Concession. It seems to have catered to those involved in the maritime life of the port:

pilots, ships' captains, and travelling salesmen, the latter commonly advertising their presence in Hiōgo in order to invite would-be buyers to inspect their lines of furniture or other specialized goods.

A number of the more established trading *hongs* were able to maintain a presence in the Concession for some duration. Smith, Baker & Co., the American firm, which also had a strong presence in Yokohama, occupied the same location on the Bund from the opening of the port to the end of the century. Others, such as L. Kniffler & Co., a German firm based in Düsseldorf with operations in Yokohama, operated at lot #12 on the Bund, adjacent to the customs house. But this firm, like others in this category, also experienced a degree of transience, reorganizing under a different name as the principals in the firm gave way to successors, most of whom had been employees. Accordingly, Kniffler had become Illies & Co. by 1888, having been taken over by Christian Illies, who appears to have been the firm's Kōbe agent from the time it opened in 1869. The local newspapers routinely provided notices of these reorganizations. Typical of these was that of the partnership between H.J. Hunt and F. Hellyer, who operated as the local agents for Alt & Co. of Nagasaki. Hunt gave notice that the partnership would end at the end of 1881 in both ports, but that henceforth two new firms to be called Hunt & Co. and Hellyer & Co. would be established at both ports.[27] Both Hellyer & Co. and Hunt & Co. would endure to the end of the treaty port era.

Improvements in the Native Town

At the time of the land sales in the Concession, many potential buyers recognized that lots adjacent to the Native Town were less desirable because of an "evil smelling drain" that ran alongside it.[28] Compared to the Euro-American landscape expected to be developed in the Concession, the Native Town was a dense warren of existing Japanese buildings and land uses that many foreigners would have regarded as an undesirable neighbour. Indeed one writer, long after the fact, likened sections of this neighbourhood to "the worst slums of Liverpool or London."[29] It is perhaps not surprising that from an early date the popular name used by the Euro-American community for Nishi-machi, the street that separated the Concession from the Native Town, was Division Street. The point where the two settlements abutted presented a sharp divide in the cultural landscape even after the tensions and mutual suspicions of the two populations had dissipated.

The presence of the Concession, in which the foreigners immediately set about fashioning an orderly Western settlement, presented the Japanese with a living laboratory of Western urban structure and architectural design. Not surprisingly, the enduring presence of a number of foreign businesses in the Native Town and the natural to and fro of foreigners had an effect on developments within this zone; for example, it spurred the construction of Western-styled buildings and other civic improvements that better dovetailed with the street pattern of the adjacent Concession. As early as 1868 the *Hiōgo News* reported that a few new houses had been built there and a great many of the old ones brushed up. It is here principally that Japanese businesses and Chinese stores and grog shops flourished:

> Nearly every street in the Native Town seems alive with carpenters and the amount of alterations made and new buildings erected during the past twelve months is very considerable ... [replacing] the thatched huts of fishermen [that once] stood on the site of what is now known as Native Bund, west of the Settlement.[30]

The Japanese authorities seem to have promoted these changes through their own building activities, such as the construction of the *Saibansho* (District High Court) in 1868 just north of the Native Town, a move that led in time to the erection of other institutional buildings nearby, including the Japanese Normal School.[31] Another major change came when the Japanese authorities significantly reconfigured the Native Town by inserting a new street – Sakaye-machi, terminating at Division Street (Nishi-machi) – on the western end of the Concession. This street opened up the Native Town and provided it with a modern street of the same proportions and style of sidewalks as those in the Concession. Among the early occupants of this thoroughfare were some Japanese banks and merchant premises, but to many of the foreigners, it remained an area distinct in look and feel from the Concession itself.[32]

This energy and change came to typify the Native Town in later decades, and the authorities responsible for it would at times demonstrate a level of "progress" that outdid their neighbours in the Concession. An example of this was in early 1884, when, as an experiment, electric street lights were installed on the Native Bund. To be sure, the experiment was modest: six lamps, each about 2,000 candlepower, with a dynamo situated in a rice mill at the Shindin (or Eastern) end of the Bund. While this initiative seems not to have been an immediate success, owing no doubt to technical problems of reliability, it did pre-date efforts to electrify the street lamps of the Concession by four years.[33] This latter step, when it came, was accompanied by a more general attempt to bring domestic electrification to the Concession. In September 1888 the *Hiōgo News* reported that the Kōbe Electric Light Co.'s test had been very successful; of the seventy buildings involved, only one (a hotel on the Bund) was unsatisfactory in its lighting, and this was thought due to interference by guests.[34] A month later the newspaper reported that steps were under way to place forty-three street lights on Sakaye-machi in the Native Town and that the Japanese authorities were seeking leave to run their wire into the Concession.[35] This proposal precipitated a heated debate on the Municipal Council, with some members advocating that the electrical wires be placed underground – a move that Kōbe Electric resisted, citing excessive costs.[36]

In 1888 the Japanese authorities in Hiōgo began work aimed at improving the drains in the Native Town. This work, however, was halted in November because all of the money allocated had been used. It was expected to resume the following April, with the hope that the job might be completed in three years. In addition, surplus tax revenue was directed toward widening Moto-machi Street[37] from the Ujikawa (a small brook that marked the western edge of the Native Town) to the first rail crossing, and similar work was undertaken on some other streets in Hiōgo City, which lay farther to the west on the other side of the Ujikawa. This flurry of infrastructural enhancement also included the construction of drains and the development of a municipal water system for the city. In the course of this project, salt water from the sea had been permitted to flow back up the drains at high tide. As a result, some wells in the vicinity of Sakaye-machi and Kaigan-dori became brackish. The foreigners living in the Concession adjacent to the Native Town evidently experienced the impact of this problem and eagerly awaited the completion of the water system, not only because it promised to alleviate a temporary inconvenience, but also because it offered the prospect of much improved public hygiene in the Native Town.[38]

The net effects of these changes in the Native Town demonstrated the eager and progressive efforts of the Japanese authorities – and indeed entrepreneurs of all nationalities living in that "sub-community" – to create a "modern" townscape as befitted the spirit and ambitions of

Meiji Japan. These physical transformations seem to have been deliberate attempts to harmonize the Native Town with the Concession, and by the end of the century, the differences between the two neighbourhoods were becoming less and less perceptible.

Fashioning a Civic Landscape within the Concession

Kōbe's late start as a foreign settlement, and the determination of its founders and early Municipal Council to create a place of order and taste – a model town – encouraged a landscape deliberately tailored to reflect these attributes. Shanghai, one of the closest and most powerful models of such a place, served as an example for Kōbe.[39] The British Settlement at Shanghai had been founded on the west bank of the Huangpu River in a marshy area to the north of the Chinese walled city. Distinct American and French Settlements were later added, and in 1863 the British and American Settlements were effectively merged. The former was initially concentrated along the Huangpu river front, which, under the direction of the British Consul, was divided into lots. Here Western business firms erected compounds containing their godowns, offices, and residential quarters. In accordance with the Land Regulations of 1845, the foreshore, along which ran a towing path for imperial Chinese grain junks, was left open to accommodate a public thoroughfare. As time went on, the occupants of the river front lots extended the foreshore to create a broader roadway, which was named "The Bund."

Among the other features of the Shanghai Settlement was a Public Garden, which was founded in 1864 when the foreshore opposite the British Consulate at the northern end of the Shanghai Bund was ceded to the settlement by Great Britain on the condition that the reclaimed land be used for this purpose. Significantly, this became a reserved park, from which the Chinese public would remain excluded until 1928. In Kōbe, as in Shanghai, there was a tacit understanding that the Foreign Concession and its Bund would be off-limits to the local Asian population. In Kōbe the municipal authorities very early on set about planting trees along most main streets in the Concession and particularly along the shore front, where a wide swathe of lawn ensured that the Bund took on the appearance of a fashionable British seaside resort. Here it became the habit for residents to gather on the promenade, where they could enjoy an unobstructed view of the harbour and presumably pass the time undisturbed by the Chinese or Japanese, although there is no evidence that Kōbe ever formally resorted to the type of racially inspired exclusion that typified Shanghai. Here too were benches where residents could sit while being entertained by a local brass band under the direction of a Portuguese resident.[40] Kōbe's Bund promenade, then, was a key public space; but there were other locations offering further amenities, such as the park associated with the Municipal Hall at the corner of Division Street and Maye-machi, and the Recreation Ground located on the reclaimed ground of the former Ikuta river bed.

Just as the planning template owed much to the norms that had been established in other treaty ports and Western settlements in South and East Asia, the same might be said of architectural norms. In Shanghai, as in the other Chinese treaty ports, the earliest buildings erected by Westerners reflected a hybrid architecture, one that echoed British neoclassicism as had been adapted to Madras, Calcutta, Singapore, and Penang. The buildings were constructed by the Chinese, who used local materials and techniques but followed designs and specifications suggested by the foreign firms. Initially, these buildings were two storeys in height; by the 1860s, several were three storeys. These buildings were notable for their thick walls, wide verandas and loggias, and tile roofs. Heubner cites a guide book of 1867 that describes Shanghai's treaty port architecture as "a style of mingled solidity and elegance which entitled Shanghai to contest with Calcutta the designation of the 'city of palaces.'" Most of the new buildings resembled Italian villas, "the classical lines of which have been strengthened into massive proportions, and modified although not obliterated by the addition of the indispensable verandah."[41] By this time, Shanghai had a resident British architect, Henry Lester, who is credited with helping merge an academic style with the practical necessities of those who commissioned these buildings.

A handful of dwellings erected by foreigners in Kōbe have survived to the present day, but most of them were constructed after 1900 and thus post-date the treaty port era. Most of them are in the Kitano-cho district of the modern city, near the upper elevation limit of urban development, just below the Rokkō Mountains. This district is on the "Hill," some distance from the original Foreign Concession and thus from the settlement on which this study focuses. These dwellings have a particular fascination for Japanese today because of their exotic Western appearance. They are also among the few foreigners' dwellings of any age to survive in the greater metropolitan region of Kōbe.

Urban housing tends to be rather transitory in modern Japanese cities, where growth and expansion are constant. The impact of the Allied

Figure 6.1. Foreigners' houses preserved in Kobe. Photo (a) (top) is one of the surviving foreigners' houses in the Kitano neighbourhood. Photo (b) (bottom) is a house relocated and preserved on Port Island, a man-made island formed during the 1970s. Both dwellings exemplify the types of wooden clapboard houses built by foreigners.
Source: Author.

bombing of industrial centres at the end of the Second World War, and of land economics that place greater value on ground rents than on built assets, has meant that few houses have lasted more than two generations over the past century. Given that foreigners tended to have different attitudes, tastes, and norms with respect to housing, it may not be surprising that the surviving older houses in Kōbe were created by the foreign community there. Several of the foreign houses in Kitano have been converted into tea rooms and restaurants because of their foreign and romantic atmosphere, and collectively they have been used as film sets when the Meiji period is being evoked.[42] Figures 6.1a & 6.1b offer examples of Kōbe's efforts to preserve foreigners' houses of the treaty port era.

As reference points for the residential landscape created by foreigners in the settlement itself, these surviving houses can be misleading. Some of them were greatly different in architectural inspiration from those that were constructed in the settlement itself, or even from those dwellings built on the Hill above the Concession during the treaty port era. Happily, there is a good photographic and graphic arts record from which to reconstruct the built landscape of the Concession during the treaty port era, and from that record, the Kōbe City Museum has produced a remarkably executed and, it would seem, authentic scale model of the Concession as it was at the end of that era. This has been very helpful in allowing the modern viewer to capture the essential appearance and layout of the Concession. Unfortunately, this representation does not extend to the Hill, where, as we will see, a rather different landscape emerged. What is clear is that the Concession evolved into a tightly configured urban townsite, albeit a very small one.

As noted in chapter 2, the lots forming the cadastre were generous in scale compared to those available to the Native Town's occupants (and to occupants of Japanese cities in general), but they were also sufficiently limited in size as to require a relatively compact townscape. The general lack of alternative expanses of cheap land, the beach locations beyond the Concession notwithstanding, required that land use be carefully structured and integrated. As will be noted below, the generously proportioned edifices of the leading merchant firms on the Bund and of the hotels and banks on Kyo-machi lent a kind of big-city impression and classicism to the urban tableau. However, the side and back streets were much more like provincial European and American mercantile neighbourhoods, where houses were at best semi-detached but more often than not formed continuous rows of modest two-storey

Figures 6.2. & 6.3. Views of the foreign settlement at Kōbe, 1878, by C.B. Bernard.
Source: *The Retrospective of Kobe Modernism: Exhibition Catalogue.* Kobe: Kōbe City Museum,1986, 12. Reproduced with permission of the Kōbe City Museum.

dwellings abutting the thoroughfare. One of the best representations of these streets is the series of watercolour paintings attributed to C.B. Bernard ca. 1878. Bernard's paintings emphasize the broad avenues and neatly ordered streetscapes of the Concession. But more remarkably, these streets seem virtually bereft of street life, save for a thin scatter of *jinrikisha* drivers and Chinese clerks and coolies. Whether this was illustrative licence on Bernard's part or an authentic rendering of the scene, one can be certain that the same would not be the case a few blocks away in the teeming Native Town.

Real estate advertising for the period offers some glimpses of the modest scale of these dwellings and the costs of acquiring them. Advertisements repeatedly refer to two-storeyed dwellings with between four and six rooms for sale or for let. Sometimes the advertisements provide evidence of other assets associated with the property, for example:

> To let: with possession 1 Sept. Commodious dwelling on lot 112 Concession containing 5 rooms with verandah facing south and west together with gas fixtures, fenders and fire irons, curtain poles and rings, force pump leading to bath-room, and flowers and plants in conservatory. [contact] J.M. Mur.[43]

A great many of these buildings must have combined housing and business premises in some fashion, and it appears that within the Concession itself, few streets or blocks faced more restricted zoning. It is clear too that many of these houses provided accommodation for a number of unrelated and largely male residents, suggesting that those buildings not maintained by one of the merchant firms for their employees, operated as lodging houses.[44]

A closer examination of Kōbe's architectural development reveals that it bears some resemblance to the British Settlement at Shanghai. Buildings within the Concession itself were typically two-storey structures constructed initially of wood and capped by a Japanese pan tile roof. Invariably these buildings were rather rectangular and block-like in their proportions, although they were less heavy in their massing and appearance than buildings in Shanghai. One reason for this was the use of wood as an economical material for smaller dwellings and bungalows, many of which featured clapboard and a full array of the decorative trim that was becoming commonplace in this period, particularly in New England and other parts of North America.[45] When painted in vivid and contrasting colours, the resulting dwellings conveyed an airy and even fanciful appearance, made more apparent by the almost universal adoption of shutters, French doors, and large, often two-storey verandas.

Few of the buildings in the Concession had wings or other appendages extending from them, other than the almost pervasive two-storey galleries. This general scaling of buildings no doubt reflected the confining nature of the lots, which demanded that buildings in the compound fit effectively with other structures that primarily served the economic enterprise of the owner (e.g., godowns, tea firing factories, and the like). The form and proportion of the buildings on the Bund suited a classical academic architectural vocabulary, and this was particularly so for those erected for the major trading firms, for banks, consulates, and hotels, and for institutions. One writer has described this tendency as an "almost excessive fondness for Ionic and Corinthian capitals, Palladian arches, pilasters, dentils, and broken pediments."[46] These choices are interesting in that from the 1870s to the 1890s, other architectural fashions were coming into vogue in the home countries of these foreign residents. Their continuity in places like Kōbe suggests either a detachment from, or a lack of knowledge of, the developing architectural modes appearing elsewhere; that, or a conservative yearning to reproduce a familiar and therefore comforting landscape as a reminder of home. According to one writer, this neoclassicism had become the architectural norm along the string of British trading outposts from India to China, and as such, it enjoyed a currency even after it had passed out of fashion in the home countries.[47]

Few of these buildings have been well documented with respect to who might have served as the architect, if indeed one was employed. We do know that after the port opened, several individuals advertised themselves as architects. The *Hiōgo News* of 1 September 1869 carried a notice that Storey and Smedley, Architects and Civil Engineers, were turning over their architectural business to the redoubtable J.W. Hart.[48] Hart was, of course, a British naval-trained civil engineer; it is not clear whether he had formal architectural training or whether he successfully developed an architectural clientele in the port. We do know that he was not alone in this "profession," for a few months later, in January 1870, the newspaper carried an advertisement for "Bonger Bros., architects, surveyors." Significantly, W.C. Bonger, a Dutch national, remained in the port, where he advertised himself as "Architect and Surveyor" as late as 1888.[49] In that same year another advertisement indicated the presence of "Wm. E. Lippert, Civil Engineer and Architect."[50] Later the port provided work for Alexander Nelson Hansell, a British-trained architect and the son of an Anglican missionary in Japan.[51] Hansell, a

The Development of a Pleasing Townscape 133

Figure 6.4. A view of the Bund from the corner of Naniwa-machi sometime in the 1870s. The building on the right at No. 10 is Gutschow & Co.; next to it, at No. 9, is the British Consulate. Note the architectural embellishments of these buildings, with their elaborate two-storey galleries overlooking the harbour.
Source: Photo 2796, used with permission of Nagasaki University Library.

Fellow of the Royal Institute of British Architects, arrived on the scene at the end of the treaty port era, at a time when many of the first generation of buildings and houses in Kōbe were being replaced by more fashionable and durable brick structures. There seems little doubt that he received a number of commissions both from the foreign community and Japanese officials. Yet there is little to tell us whether these men received many or few architectural commissions or whether, instead, their principal work involved surveying and carrying out civil engineering works under contract to the Japanese authorities. However, there is cursory evidence that a few buildings in the Concession carried a very "Dutch" appearance outwardly. The business premises of Delacamp & Co. at lots #121 and #122, for example, carried the very characteristic curvilinear concave and convex gable end profile associated with the seventeenth- and eighteenth-century Dutch building tradition. At least

Figure 6.5. Looking north on Akashi-machi from the Bund, ca. 1870. The colonnaded building on the right is No. 5, the Netherlands Trading Society. On the left on lot #4 is the godown of Smith, Baker & Co. The pine grove north of it is the Public Garden. The two-storey building a little north of the park appears to be the Municipal Building.
Source: Photo 1227, used with permission of Nagasaki University Library.

Figure 6.6. The Walsh, Hall & Co. premises, lot #2, the Bund, constructed ca. 1870. This was the most architecturally imposing of the trading company establishments.
Source: Harold S. Williams Collection at the National Library of Australia, ACT Australia.

Figure 6.7. Motomachi Street in the Native Town. This photo looks westward from Division Street toward the entrance of Moto-machi Street and the Native Town. The building on the right is the Kōbe Bazaar, a genre of store that gained popularity in Kōbe beginning in 1883. The following year the number of bazaars grew to thirteen. Immediately adjacent to the foreign settlement, many of the shops catered to foreigners and English signs became commonplace.
Source: Photo 2819, used with permission of Nagasaki University Library.

Figure 6.8. The "Native Bund" showing the Hiōgo Hotel, which occupied the first lot immediately west of the boundary of the Foreign Concession. The grassy expanse in the foreground is the promenade that occupied the ground between the Bund (Kaigan-dori) and the harbour sea wall. The roadway with the pavement roller was the junction of the Bund and Nishi-machi, also called Division Street because it divided the Concession from the Native Town to the west. Out of the picture to the left was the Meriken Hatoba ("American Wharf").
Source: Photo 1227, used with permission of Nagasaki University Library.

Figure 6.9. A view from within a compound in the Concession showing the business premises and godown (left) with its iron security shutters. This photo was taken in 1934, but reveals buildings that probably date to the 1880s.
Source: Harold S. Williams Collection at the National Library of Australia, ACT Australia.

one other building in the Concession, at lot #57, exhibited this same profile. What is nevertheless evident is that the architectural character of the Concession exuded a sense of place and coherence. A few examples will help show this.

Foreign Occupation of "the Hill"

The Concession proved to be hot and dusty in summer, and many residents began to seek building sites on higher ground. Within fourteen months of Kōbe being opened, a number of foreigners had procured residential sites on the Hill, where they set about erecting what were commonly described as bungalows.[52] One of the earliest to do so was Mr Korthalls, head of the Netherlands Trading Society, besides being the Dutch Consul. Korthalls built a bungalow and extensive gardens along what became Shimo Yamate-dori, the first major east–west road

The Development of a Pleasing Townscape 137

Figure 6.10. A panoramic view of the Concession looking northward, ca. 1880, revealing the mountain that backed the settlement at Kōbe. The street running diagonally in the foreground is Naniwa-machi. The building in the centre left with the two-storeyed verandas is the Hôtel des Colonies at No. 56. Note the presence of godowns in this area of the settlement and the paucity of more stylish buildings. Also visible is a line crossing the centre of the photo from east to west. This is the railway, which began operations between Kōbe and Ōsaka in 1874. The grove behind it is the Ikuta Shrine. In the background is the scatter of Western occupied houses on the "Hill."
Source: Photo 2793, use with permission of Nagasaki University Library

to traverse the foothills above the Concession. Gradually other foreign-owned dwellings with large gardens crept higher up the Hill.

Thus the foreign residential areas of Yamamato-dori and Kitano-cho came into being. Soon the homes and gardens on the Hill began extending westward toward Suwa-yama, or Venus Hill as it came to be called by the foreigners, because it was from there that a party of French astronomers observed the transit of Venus in 1874.[53] Foreign acquisition of housing was not limited to the Concession itself, however. Once it was confirmed that the Japanese authorities would allow foreign occupation of houses beyond the Concession – albeit under terms different from those applying within the Concession – there was an active effort to entice foreigners to acquire these properties for speculative and investment purposes. The following advertisement in the *Hiōgo News* of 9 March 1870 read:

> Real Estate Sale – Wainwright & Co. Auction on Thursday March 10th. A plot of ground containing 308 tsubos lying immediately north of the adjoining Foreign Concession, with a row of 5 houses standing thereon, commencing from the Guard house on Temple Road ... leased for the Japanese Government at annual rent of 308 boos payable quarterly in advance subject to those taxes as conforms to the Treaty and Convention Stipulations for Hiōgo. Public Road 50 feet wide runs both in front and back ... Houses are Japanese built, two occupied by tenants ... must be sold as one lot.

No doubt, the proximity of this location to the Concession and the presence of already existing housing was an inducement to would-be purchasers. For some this activity was a calculated speculation in anticipation that the railway would have to buy them out. But for many the attraction of locating on the Hill was that they would be farther away from the Concession, and before long a number of foreigners had established themselves in this more rural environment. Some achieved access to these locations through the agency of their Japanese wives. The *Hiōgo News* reported that those who opted for the Hill found early on that the Japanese practice of cremating dead bodies in the open and the resulting stench carried by the wind intruded on their locations. This practice was soon curtailed by the authorities after complaints were made.[54] Isabella Bird, writing in 1873, would characterize this neighbourhood as consisting of "a number of foreign wooden houses struggl[ing] up the foothills at the back [of Kōbe], some of them unmistakably English bungalows, while those which look like Massachusetts homesteads are occupied by American missionaries."[55]

Many foreigners were living on the Hill by 1880, but it was only in that year that an effort was made to construct a proper road to it. This was completed in 1888, by which time the pace of residential expansion in this zone had accelerated. Undoubtedly this type of housing held considerable appeal for those able to indulge in it. The elevated locations offered a slight respite from the sultry conditions of the summer months as well as some separation of home from the dense press of humanity in the Japanese settlement below. There were concerns over sanitation, and cholera was a seasonal dread for foreigners – not without reason, for the port was repeatedly visited by cholera during this period.[56] That said, the prospect of having a garden must have motivated many to choose the Hill. Gardening was an accepted genteel pursuit for Euro-Americans in the latter half of the nineteenth century, and for those who lived abroad in exotic habitats, it offered the prospect of exploring new species and horticultural designs.

The elevated slopes of the Hill and the Rokkō Mountains, which Rutherford Alcock, the first British Minister to Japan, had described as forested in 1867, were largely denuded of forest cover during this period. It is not clear why and by what agency this happened. The demand for wood to construct the burgeoning Japanese city of Hiōgo must have played a part in this, but so too perhaps did the demand for charcoal, the principal fuel source for heating and cooking among the Japanese. Needless to say, the stripping of forest cover had its own consequences with respect to erosion, mud slides, flooding, and the accelerated run-off of rainwater, so that by the end of the nineteenth century, local authorities sought to reverse the process through reforestation.[57] The presence of foreign occupants on the Hill only reinforced the impression, for locals and visitors alike, that the treaty port was unique among its counterparts in the Far East.

Industrial Land Use on the Edges of the Concession

If the Concession itself presented an orderly and pleasing townscape in the eyes of many contemporary observers, and the Hill an enchanting and bucolic residential haven from the town and city at its foot, the eastern and western margins beyond the Concession presented a somewhat chaotic zone. As soon as Kōbe opened to foreigners, a distinct economic geography emerged, one that resulted in part from how the Foreign Powers envisioned the Concession and in part from the ways that individual actors sought out particular settings for their activities. For example, "dirty" industrial activities – foundries, machine shops, shipyards, construction and assembly yards, and the like – found no place within the Concession. Instead, they located along the beaches east and west of the Concession, where they enjoyed proximity to the harbour for landing raw materials and for hauling and launching vessels. Tea firing, however, was conducted within the Concession. This activity had it critics, who viewed it primarily through the lens of social disapproval: residents were offended by the noisy crowds of Japanese workers, who broke the sanctity of the Concession to conduct that work. Presumably, the sweet-smelling smoke of the ovens could be tolerated, for those ovens were the port's raison d'être and their smell was of money. Nevertheless, this clear segregation of industrial activities meant that the built landscape of the Bund in the Native Town was far different from that of the Bund in the Concession. The former was quasi-industrial and ragged, as was characteristic of such streets; it would have been lined with piles of timber, coke, and coal, half-built ships, and an assortment of cranes and shipyard "ways." It would also have provided a setting where Japanese businesses and residences stood side by side with those of Westerners – something that was prevented within the Concession by the land rent arrangement worked out between the Foreign Powers and the Meiji government.

Imagine taking a walk westward along the Bund in about 1872 from a point in front of the Cornes & Co. premises and adjacent to the Merikan Hatoba at the foot of Nishi-machi. One would first encounter the store and godown of J.D. Carroll & Co., a ship's chandler and storekeeper. Next was a small house occupied by Mr Lake, then a little bungalow occupied by Textor & Co., a German firm whose iron godown stood on the adjoining lot. Farther along, one would come to the workshop and residence of Fitzgerald and Strome, a shipbuilding enterprise, and beyond that a Japanese godown and three or four Japanese houses. One would now have reached the Western Customs House, which occupied a sizeable tract adjacent to the harbour. Continuing westward there would be more Japanese houses, the shop and residence of Mr Frey, a shipbuilder, and farther along the beach a series of significant proto-industrial fabricators beginning with the Kaga Foundry, the Vulcan Foundry, Glover & Co.'s sawmill, the Wignall Ship Yard, and the Fitzgerald and Strome Ship Yard. Next were the Frey Yard, Thayer & Co.'s Yard, and Board & Co.'s Yard. At this point the observer would have walked about halfway around the sweeping bay fronting the Native Town to the outlet of a small creek called the Uji River and would be looking out toward the point at Benten no hama, so called because it was near the Benten Shrine on the east bank of the Uji. Here was located a dockyard under American management. Beyond loomed the Japanese city of Hiōgo.

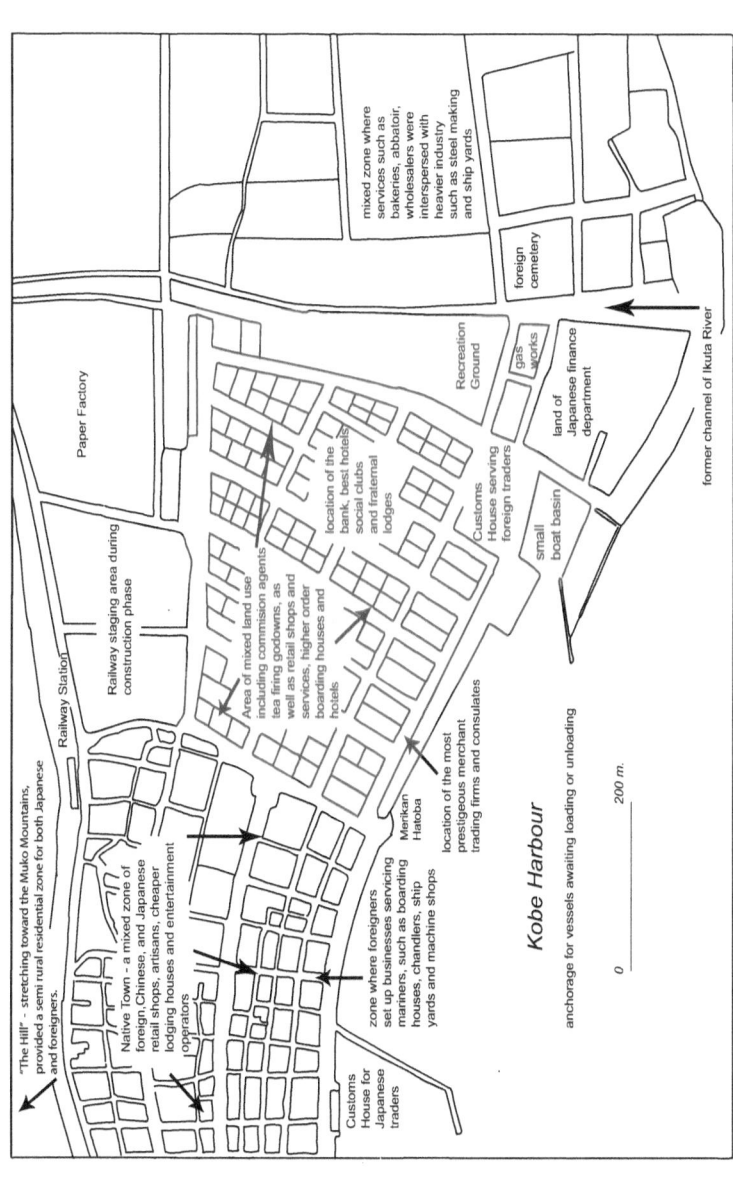

Map 6.3. A map of functional land use in and around the Foreign Concession, ca. 1880.
Source: Author.

Based on the details provided by the *Hiōgo News* in 1869, a similar walk eastward from the Concession along the shore would have presented one with a similarly diverse set of land uses. This zone lay to the east of the Ikuta River, and the road traversed was commonly called Water Street – a term often applied to such streets in the port towns of colonial North America.[58] The first premises encountered would have been those of Kirby & Co.'s abattoir and cattle sheds, and perhaps, incongruously, a "neat bungalow" (presumably for the employee overseeing this operation). Next were the butcher shops of Adams & Co. and Warren Tillson & Co., both of which also operated bakeries. Adjacent to these was the Hiōgo Hotel,[59] then the store and bakery of Kirby & Co. and two rows of dwellings facing the beach, which were undoubtedly rented out to occupants by their owner, Carroll & Co. Beyond these were Mr Nethersole's new two-storey house, which served as a billiard parlour; then another new two-storey house, with livery stable, owned by Rangan & Co.; and then a bungalow, which would soon be removed by Adams & Co. to make way for their bakery. Finally there was a building belonging to Hobson & Co., which according to the report was undergoing extensive alterations for it to serve as premises for a social club. The upper rooms were to be a reading room and reception area, the lower room for billiards; also, a substantial veranda was being constructed the entire length of the building facing the street. It seems likely that this reference was to the future premises of the German Club. Within the year, further building projects along this stretch of Water Street would be reported, including the following: additional improvements to Kirby & Co.'s and Adams & Co.'s premises; a small house and godown for Mr Grosser; several Chinese stores and residences; and godowns for Hecht Lilienthal & Co., Wachtel Groos & Co., Walsh & Co., and the Netherlands Trading Society (the latter two firms also held prime lots on the Bund within the Concession). Also, new houses would be built to serve Case & Co.[60]

Land immediately to the north of the Concession was employed in 1870 to satisfy other service and residential needs of the newcomers. According to the *Hiōgo News*, the buildings here included the following: a fine, European-built house, which served as offices and residences for the Netherlands Trading Society; on the west side, a moderate-sized house occupied by Richter and Reinhardt; billiard rooms and a bowling alley for N. Mancini; offices and a residence for Schultze, Reis & Co.; and a godown and residence for Wainwright & Co.[61]

Figure 6.11. A diagram of the spatial model of Kōbe as a mercantile settlement.

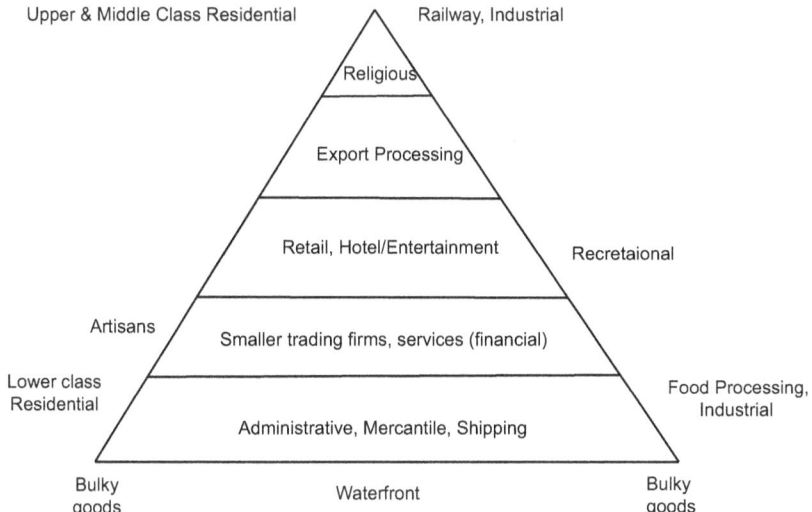

Source: Author.

In the years that followed, these *de facto* appendages of the settlement evolved into more clearly "industrial" landscapes. For example, the section immediately to the north of the Concession would be the site for the Walsh brothers' Kōbe Paper Factory in 1877 as well as the Sannomiya railway station and rail line. Similarly, with the relocation of the Ikuta River, land use to the east adjusted accordingly, primarily to accommodate the growing operations of Kirby & Co., which in the mid-1870s moved from being primarily a provisioner to become a major shipbuilding and iron-making enterprise, the Kōbe Iron Works. This operation absorbed a large tract along Water Street. The gasworks would also be located in this sector of the town, as well as a large storage and marshalling yard associated with the Eastern Customs House, operated by the Japanese Department of Finance. Buffering the Concession and these utilitarian land uses was the Recreation Ground, a large wedge of land immediately to the east that had been part of the course of the Ikuta River. This plot was an important amenity for the foreign community, particularly given the port's male-dominated society.

These observations take us back to the models for colonial port cities that were presented in chapter 1. Bowden's model (See Figure 1.1) asserted that land rents determined specific land uses within a settlement and invariably resulted in a triangle of distinct activities.

Figure 6.11 extrapolates from Bowden. It is clear that at Kōbe, the waterfront land along the Bund was of the highest value; accordingly, access to it was commanded by importers and exporters of high-value goods. Wharfs and related maritime activities such as shipbuilding were relegated to the waterfront's periphery, where land was cheaper and where these activities would not conflict with expectations that Kōbe's Foreign Concession would be a townscape of commercial substance – one that would house a middle class as well as institutions to make their social life bearable. The central section of the waterfront, therefore, was dominated by commission agents and shipping offices and their warehouses, but also by merchant exchanges, banks, and insurance offices. Farther inland were two zones that provided services to support the resident and transient mercantile populations. Here were located whatever higher-order retail shops the community supported, and also hotels and social and recreational clubs, although these were initially relegated to the zone immediately outside the Concession. It is important to note that these "leisure" activities were divided along class lines. Those catering to sailors and other working-class men were outside the Concession throughout the period; those catering to traders and other middle-class occupations – mainly social clubs and fraternal organizations – found opportunities to locate within the Concession. Football and cricket clubs necessarily found open land for their activities on the periphery, where the race course was also located.

At the apex of Bowden's merchant triangle was a zone where political and religious affairs were situated, beyond which there might be upper-class residential areas. There were two small churches within the bounds of the Concession; many contemporary observers, however, regarded these as having a weak presence in Kōbe. Nor was a strong sense of civil government associated with the settlement. This was due in part to the multi-national treaty port structure, which distributed several aspects of governance – civil and criminal, and also commercial – to the consular officials of the major national groups at the port. As has been noted, municipal administration rested with a Municipal Council, which had a physical presence in the form of a Municipal Building that housed key functions: a council chamber, a jail, and offices for the town manager and the police and fire services. Consular court proceedings were frequently conducted in the Municipal Building. A closer examination of how residential structures were distributed reveals that along the backstreets and side streets of the Concession, a number of boarding houses and small dwellings were for lease; these invariably accommodated foreign clerks and other staff, who were vital to the commercial enterprise of Kōbe. This form of housing, which catered principally to single men, needs to be distinguished from the bungalows on the Hill above the Concession, where men with families, or men who had formed long- or short-term "marriages" with Japanese women, were likely to be found.

In short, Kōbe's Foreign Concession had three distinct zones. The first was the "bounded" Concession itself, in which the commerce that was the raison d'être of the treaty port was located and ordered in ways that reflected the morphology of other mercantile port cities. Within this zone were visible many of the conventions of civic space, both formal and informal. The second was the waterfront periphery, the site of "dirty" industries and nautical support services, interspersed with houses and amenities for the teeming Japanese and Chinese population that supported many of these activities. Finally, there was a suburban zone on the Hill above the Concession, which provided release and repose away from the demands of business and of imposed and regulated social conventions. In this zone, Westerners might freely mix with their Japanese neighbours and sample the exotic culture and natural environment of Japan. In this encounter one might think that the Japanese and the foreigners were more likely to meet as equals, but this is hard to gauge. It can be said, though, that this physical encounter was not replicated in many other of the mercantile cities of East Asia.

When we juxtapose the model shown in Figure 6.11 with the one developed by Bowden in Figure 1.1, we find a remarkable accord between these land use overviews. It would seem, then, that what emerged in Kōbe mirrored the morphology, or patterns of land rent or land use, found in other parts of the global mercantile world before 1900. That the example of Kōbe was executed under the strictures imposed by the Japanese underscores the mercantile habit of mind that Westerners had worked out over three centuries.

Chapter Seven

Life at the End of the World: Forming an Expatriate Society in Kōbe

Population Growth and Change

The initial population of the Foreign Settlement at Hiōgo consisted of the foreigners who poured into Hiōgo in 1868 in anticipation of the commercial opportunities the port was expected to afford. It is difficult to place a number on this original population, for many of these people undoubtedly would have been there as part of the Western powers' apparatus for securing the location through a military presence. Nor is it easy to determine with precision the population numbers thereafter, owing to the multi-national governance of the treaty ports, which meant there was neither a requirement nor a single agency to conduct a regular and systematic census in the Concession. Moreover, many of the foreigners initially lived in the Native Town and thus may not have fallen under the administrative oversight of their respective nations' consuls. The British and American Consuls did, however, provide their superiors with population figures for their respective nationals who were assumed to be connected with the settlement. These figures would in time include missionaries, and also the "foreign experts" recruited by the Japanese authorities to help build infrastructure such as lighthouses and railways. Many of these people would have spread out through the hinterland of Kōbe, beyond the Concession and the city of Hiōgo.

Apart from the consular returns, the Municipal Council constructed lists of those eligible to vote in municipal elections. These lists were typically ordered by nationality and were based on the limited franchise, which was restricted to males resident in the port who had registered with their nation's consul. Consequently these lists did not include the Chinese population in the port, nor did they include the wives and children of those making up the foreign population, nor is it possible to gain from these data any insight into the extent to which foreign residents entered into permanent or impermanent "marriages" with women of Japanese or Chinese origin – a practice that effectively increased the real dimensions of the Euro-American population, if one accepts that such families were part of the fabric of the foreign community. The exclusion of this Asian and Eurasian population from the record points to the ambiguity of this self-styled Euro-American community. Also absent from any of these records are the large numbers of transient seamen who passed through the port each year. Hoare suggests that this sub-population numbered between 4,000 and 6,000 for Kōbe in the early 1870s. They formed part of the "resident" population of the moment, so their absence from the recorded population inevitably blurs the foreign presence in the port throughout this period.

All of these caveats aside, the most reliable population data seem to be the annual figures recorded by the British Consuls at Kōbe. These were undoubtedly constructed with the cooperation of the other foreign consuls as well as the municipal and prefectural authorities. These data provided only head counts of the foreign population by nationality and therefore are an imperfect means for measuring demographic components and other qualitative details of this population. For example, they do not allow us to systematically gauge the age structure, marital status, or mortality rates of the population. The surviving data do provide some indication of the number of foreign women and children within the numbers counted, but we cannot assess how many men might have had families residing in their home country or elsewhere.

It is also the case that even the "permanent population" was very fluid, owing to the high geographic mobility that was built into the mercantile economy. For example, the sources used to reconstruct the foreign population often anecdotally refer to members of trading houses who were absent at the time of enumeration; this reflects the degree to which many in this mercantile world circulated among the treaty ports both in Japan and farther afield on the China coast, or indeed back to home countries on periodic furloughs. All of this resulted in substantial shifts from one year to the next in both the absolute population numbers themselves and in the individuals forming the population. Nevertheless, the British Consuls' population estimates provide a glimpse of the foreign community and offer a means to assess the relative balance of nationalities and the general ratio of the sexes, besides providing some sense of the settlement's transient nature. An analysis

148 Opening a Window to the West

Table 7.1. Estimates of foreign population at Hiōgo by nationality, 1870–96

	1870	1873	1876	1880	1883	1886	1890	1893	1896
American	38	24	37	49	33	52	87	101	126
British	112	210	212	216	232	223	310	351	515
Dutch	26	28	19	14	12	12	13	11	9
French	27	31	5	12	14	26	54	31	36
German	40	62	38	48	56	92	107	141	141
Portuguese	5	16	5	10	17	22	35	26	45
Other	23	44	15	22	13	13	17	32	39
total	271	415	331	361	369	404	608	651	922
% change	–	53.1	–20.2	9.0	2.2	9.4	50.5	7.1	41.6
Chinese	n/d	n/d	n/d	516	617	596	913	1004	1113
% change	–	–	–	–	19.6	–3.4	53.2	10.0	10.9

Source: British House of Commons, *Parliamentary Papers, Commercial Reports, Japan, vols. 4–10* (Shannon: Irish University Press, 1971).

of these data, however imperfect, can also be correlated to the previous discussion of the economic pulsations of the port. This analysis is also central to a discussion of the social structure and social life of the port.

Table 7.1 indicates the growth of the foreign population from 1870 to 1896 using approximately three-year time intervals. These data reveal the modulations in the growth of the population, most noticeably as a result of the economic slowdown of the mid-1870s, which resulted in a 20 per cent decline of the Euro-American population. Generally thereafter, the trend is upward, although at anything but a steady pace. The growth of the Chinese population also reveals a choppy pattern, and noticeably is not synchronous with that of the Euro-American population. What is *not* revealed by this table was the sudden collapse of the Chinese population owing to the Sino-Japanese War of 1894–5, during which most of the resident Chinese evacuated the port in response to the evident tension and danger to life and limb that surrounded their presence in the community at that time. What is perhaps remarkable is how quickly the Chinese returned after hostilities ceased. These observations aside, perhaps the most striking aspect of this picture is that, while the Chinese outnumbered the Euro-Americans for as long as data were available, with few exceptions they resided outside the Concession and rarely formed part of the collective foreign presence within it.

It is also important to place the foreign population in the context of population growth within the broader community of which it was a subset. In 1880 the American Consul reported that according to the latest prefectural census, there were more than 2,500 houses in the Native Town of Kōbe, which were inhabited by more than 18,000 people exclusive of foreigners. Also part of this population were 347 American and European residents, who inhabited 99 houses, and 545 Chinese occupying 46 houses. Here it is assumed that the Japanese were recording some part of the floating population of sailors and the like who flowed through the areas surrounding the Concession, but this census may also have captured those permanent foreign residents who lived on the Hill – individuals who were also recorded in the population numbers generated by the British Consul (see Table 7.1). Thus the prefectural figures undoubtedly involved some double counting. Nevertheless, taking these figures at face value and adding the 316 Americans and Europeans estimated to occupy the Concession, the foreign population of "greater Kōbe" in 1880 can be estimated at around 700, exclusive of Chinese.[1]

In contrast to this, the city of Hiōgo had a native population of 35,000, among which there were no reported foreigners. The Governor of Hiōgo reported that some 2,764 of these people were temporary residents who had come from surrounding districts – principally Ōsaka and Kyōto, but also the city of Ōkayama to the west. While his report was not as clear as it could have been, we can presume that by "temporary resident" he meant migrants in the process of moving to Hiōgo, since his report added that the number of houses was increasing by about a thousand a year.[2] This was impressive population growth, but it paled in comparison to that of Ōsaka, which by 1880 had become an urban conurbation of more than 600,000 people, of whom only 78 were foreigners and 112 were Chinese.[3] Clearly, the region's foreign population remained small and highly concentrated, amid a sea of Japanese, who were themselves flooding the region's large urban places as a result of the rural-to-urban migration that became possible after the Meiji government released the Japanese peasantry from the land to which they had been bound.

Nothing in the growth of the foreign population suggests a strong demographic or natural increase. Rates of natural increase were falling in Britain, Western Europe, and America in the latter decades of the nineteenth century as the demographic transition associated with the Industrial Revolution took hold. In broad terms, growth rates for

Table 7.2. Women and children in the foreign population of Hiōgo, 1878–93

	1878	1886	1890	1893
Euro-American				
males	254	213	358	389
Females	60	77	125	122
Children	71	114	125	140
Chinese				
Males	n/d	463	711	602
Females	n/d	43	112	111
Children	n/d	90	90	291

Source: *British Parliamentary Papers, Japan*, vols. 4, 5, 6, 8, 9, 10.

these "homelands" were in the range of 2.0 to 2.25 per cent per year. If the Euro-American population of Kōbe consisted of a representative sampling of this mature and family-dominated population structure, the three-year growth intervals should have produced growth of at least 6 per cent due to natural increase alone. The data shown here for the Euro-American and Chinese populations suggest that migration, both in and out, had a very considerable influence and was the dominant factor affecting rates of change within the population. Indeed, when we add the evidence of the presence of females and children in this population, it becomes obvious why natural increase was such a weak factor. As Table 7.2 shows, only about one-third of Euro-American men residing in Kōbe appear to have been accompanied by a foreign-born wife, and when we average the numbers of children within this female population, we find that family sizes seem to have been small at best. For the Chinese population, the picture is skewed even more toward domination by males, with the ratio of males to females ranging between 6:1 and 10:1. Conventional families were comparatively few within both population groups, although we can assume that most of the foreign women in Kōbe were married, given the limited economic means available to women in this setting.

Why were families so small? For the Euro-American population the limited opportunities for educating children within the community, particularly as they reached their teens, meant that children would have been sent elsewhere, or "home," to complete their schooling. In the same way, we might conjecture that although a few families exhibited conventional Western patterns of family formation and life, the social fabric was truncated with respect to those norms. Thus while some younger men and women might marry and start families in the Far

East, those who were able soon moved to locations where there were larger concentrations of their own people as well as social and educational infrastructures to sustain the conventions of late-nineteenth-century life, whether it be Yokohama, Shanghai, Hong Kong, or back in their native land. Then there were some in this population who opted for other ways of life that deviated from these conventions.

National Composition of the Population

Table 7.1 reveals that the dominant national groups were the British and the Germans, with much smaller clusters of French, American, Dutch, and Portuguese nationals. Subsumed within the "other" category were small numbers of Italians, Swiss, Belgians, Austro-Hungarians, Swedes, Danes, Russians, and, late in the period, Hawaiians, who episodically lived in the community. Also, subsumed within the "British" category were Canadians and Australians and various other colonials (from India, for example). British and German residents together comprised around three-quarters of the Euro-American residents throughout the period.

At the heart of the strong British and German presence were import–export firms. From the outset, British firms outnumbered those of other foreign nationals. In 1871 the British Consul recorded that of the 103 businesses at Hiōgo, 43 were British-owned. The next three leading national interests were the Americans, with 17 businesses, and the French and Germans with 14 and 13 respectively. It must be noted of course that the businesses referred to ranged from merchant *hongs* and banks at one end of the scale to billiard parlours and grog shops at the other, with a variety of enterprises – including ironworks, marine pilots, churches, doctors, and chemists – somewhere in between. Undoubtedly the British and other leading foreign groups were more diverse by virtue of their greater numbers and the complexity of the operations they brought to Kōbe. Thus a British merchant *hong* might employ from two to five of its own nationals, some of whom might be accompanied by a wife and children.

Perhaps surprising is the limited American presence. It is clear that American interests were more focused on Yokohama, where they were second only to the British in numbers and influence. It is also the case that American traders were still preoccupied with China, where the major American overseas trading firms of Augustine Heard & Co. and Russell & Co. were powerful forces in many ports. The strength and influence of Heard & Co. was beginning to wane, however; it had established a presence in Kōbe by the mid-1870s, but it collapsed in bankruptcy soon after. Heard & Co, Walsh Hall & Co., and Smith, Baker & Co. were present in the port, but no other major American trading companies, and the few Americans who did settle in Kōbe participated in a variety of other service and entrepreneurial activities. Except for the tea trade, which so dominated Kōbe, the American presence might have been even more muted. Nevertheless, Kōbe saw its share of American mariners, particularly sea captains and their officers, who were prominent in the carrying trade. These men, however, were part of the port's impermanent population and did not establish residency or connections with the community in the way that land-based merchant traders and the commercial petit bourgeois might. They were, as a consequence, an underrepresented element of the official population snapshots created during the period. Missionaries were an important subset of the American presence in the port, and this group deserves investigation if for no other reason than they seem to have stood apart socially from the rest of the foreign community.

Death and Disease in the Port

From the outset of the settlement at Kōbe, there was concern about epidemic diseases entering the port. As noted in chapter 2, both the Japanese and the Foreign Powers were quick to establish hospitals adjacent to the Concession, expecting that these could serve to quarantine anyone believed to be carrying infection. It was also noted that those responsible for preparing the Concession site were at pains to address issues of public hygiene by constructing and maintaining drains and by establishing a "scavenger corps" to keep the streets clean. There is evidence too that the Japanese authorities attempted to address issues of public hygiene in the Native Town. The *Hiōgo News* reported in 1878 that the prefectural authorities were erecting "40 brick closets" [toilets] in the city; the editor regretted that in each instance the open side was to the front, whence passers-by gained the best view of proceedings within.[4] Consuls were regularly advised of disease outbreaks in Chinese and other Japanese ports and were expected to monitor these threats as they arose.

In the fall of 1877, for example, there were reports of cholera and smallpox on the China coast.[5] Another report, in 1885, warned that cholera has been reported at Nagasaki.[6] By October of that year the disease had entered Hiōgo and Kōbe, but this seems to have been a comparatively minor episode.[7] Another outbreak occurred the following May, and yet another in August 1890. This latter outbreak proved more deadly: there were 1,365 cases in August alone, and 125 new cases and 85 deaths in early September. A similarly deadly outbreak occurred in July 1895, when there were 104 cases at Kōbe; at Ōsaka in the week ending 18 July, there were 805 cases of cholera and 730 deaths. This episode raged for weeks and only ended in November of that year.[8]

In each of these cholera outbreaks, most of the victims seem to have been Japanese, although the American Consul provided details of American nationals who had been struck down. And cholera was not the only scourge felt in the port. The *Hiōgo News* provided a statistical summary for 1881 of diseases in the broader community as compiled by the Hiōgo Prefectural Board of Health:[9]

73 cases of cholera	28 deaths resulted
848 cases of typhus fever	298 ditto
21 dysentery	0 ditto
73 diphtheria	9 ditto
24 smallpox	8 ditto

Clearly, typhus was a serious problem, made more challenging no doubt by the rapid growth of the Japanese population in the city of Hiōgo. So too was smallpox. The American Consul cited a report from Japanese officials in January 1893 of such an epidemic in which 319 new cases and 141 deaths had occurred in one week, after 221 new cases and 108 deaths the previous week.[10] Typhus was spread to humans by lice from hosts such as rats. Dense human settlements where there was insufficient attention to good public hygiene allowed the spread of the carrier and challenged authorities, particularly given that the seaborne commerce of the port brought a constantly changing population of sailors and others into contact with the city's burgeoning population.

That said, the low mortality figures for foreign residents in the Concession suggest that they were insulated from the worst of these outbreaks by their less dense living arrangements and more advanced levels of public hygiene as well as by the fact that the settlement had resident physicians and druggists. They also had access to the International Hospital, which offered care delivered by Western physicians based on four levels of clientele: 1st, 2nd, and 3rd class and charity. A snapshot of admissions and outcomes for patients at this facility in the month of August 1888 suggests the general health in the Concession. Of the seven patients in hospital during the month, one was described as 2nd class and six as 3rd class; one from each of these classes died while undergoing treatment.[11]

The American Consul recorded the deaths of American nationals associated with the community or whose ships passed through the port. Many of the latter deaths occurred at sea and involved individuals who were not American citizens, but who were still the consul's responsibility because they were crew members of American vessels. When the death was accidental or involved misadventure, the consul was required to convene a coroner's inquiry and to notify heirs and return what passed as the deceased's estate. Many of these cases involved individuals with few means, which imposed the burden of their burial on the consul or on the victim's associates.[12] When the deceased was an established member of the foreign community, the consul's record of death reveals today something of his or her life and activities.

For example, Edward M. Kuhhardt, who drowned on 5 August 1882 at the age of forty-nine, had been in Kōbe since 1875 and was agent for the Swiss firm of Messrs Favre-Brandt.[13] Kuhhardt was apparently doing a small commission business on the side, and an inventory of his estate indicates he was the owner of lot #77 in the Concession. The consul's record also refers to a godown with a large inventory of wine and spirits and other merchandise.[14] Overall, the death rate among American and other members of the Foreign Concession was remarkably low. A report to the Municipal Council in 1884 on the Foreign Cemetery records a total of 16 internments during the year: 8 residents, 6 non-residents, 2 seamen.[15] The broader picture gleaned from the consular evidence suggests that residents of the port typically died from old age or from chronic diseases such as heart disease and stroke. While the sample is small, there is nothing in these records that points to Kōbe being an unusually unhealthy place; indeed, the opposite seems more likely. But this may reflect that throughout the thirty years of the treaty port era, Kōbe attracted young adults, only a few of whom persisted in the port and would have been reaching old age as the era closed.

Domestic Arrangements and Family Composition among the Foreign Population

As noted, the foreign population was from the outset skewed toward men. The port's economy, its distance and isolation from the residents' home countries, and the lack of access to broader social networks, all promoted this imbalance. Furthermore, security concerns in the early years as well as general suspicions about residing among "heathens" did little to induce Euro-American families to reside in the port when it first opened. Only gradually did those who took up long-term residence in the port bring their wives, or form domestic arrangements that included women and children. The consequences of this imbalance in the sex ratio generated patterns of behaviour that were, not surprisingly, focused on a particular brand of male sociability that stressed participation in sports and a great deal of drinking on the one hand, and the formation of domestic arrangements involving Japanese women on the other. The bar, the gentlemen's club, and robust sporting and other forms of entertainment were the bedrock of upper-middle-class Victorian male culture.[16] By openly forming interracial "family relationships," however, these men crossed a deeply ingrained boundary of social convention.

The scholarship of the past three decades has added greatly to our understanding of the contradictions that permeated Victorian social mores. The many studies of prostitution and of male and female sexuality and eroticism in European and American societies during the second half of the nineteenth century have highlighted the extent to which middle-class male culture was conflicted about the taboos of that period. For their part, the Japanese were more open about the need to provide sexual outlets for such a male-dominated population. When the port of Yokohama opened, the Japanese authorities there made sure that a controlled, licensed quarter – called the Miyosaki or Gankiro – would be available to the foreigners.[17] In Hiōgo, a similar district would already have been in place, and it seems that with the opening of the treaty port a second such district arose in the Native Town near Sannomiya. Known locally as the Fukuhara, this facility offered access to licensed Japanese prostitutes. Hugill refers to a mariner who deserted ship in Kōbe in the 1880s and who stayed at Madame Otome's on the Kita Nagasa Dori, which ran through this district.[18] In addition to these, there were apparently at least two foreign-owned brothels catering to foreigners: Dawn's and Virginia's, named after their respective owners. Hallett Abend suggests that the proprietors openly advertised "new women" as they arrived in Kōbe as part of what seems to have been a regular tour from port to port in East Asia. Some of these women were apparently based in San Francisco's notorious "Barbary Coast." At the other end of the scale were several notorious beer halls, such as Fat Arsed Jane's, that catered to sailors and in which one could hire prostitutes.[19] Indeed, Hugill suggests that Kōbe acquired a reputation as the one of the favourite ports in East Asia for British Jack Tar because of the number and variety of tea houses, gambling joints, and brothels, many of which carried names such as the Happy Garden, the China Dog, and Back of Beyond.[20]

For most of the men who took up permanent residency in the port, prostitutes offered neither ideal nor enduring relationships. Among men who were able to acquire independent housing, it was the convention to employ a woman or houseboy as a servant to take care of domestic chores. Invariably these were Japanese or Chinese, and it seems that many such arrangements involving females led to conjugal relations that inevitably resulted in children. It is difficult to get a good measure of the extent of these unions because of the rather cryptic evidence that must be interpreted. Take the record of the death at age fifty-six of John Curtis, a resident of China and Japan for many years. The American Consul recorded that Curtis left all his belongings to his son, except for a bequest to his housekeeper, in trust for her infant son. May we interpret that the child was his, or was there another explanation for the child's paternity?[21] A few of these relationships were sufficiently open that they formed part of the historical record handed down from the period. The U.S. consular record does reveal cases of American men marrying Japanese women: George C. Foulk,[22] age thirty, born in Marietta, Pennsylvania, on 10 September 1887, married Murase Kane, age twenty-nine, from Kokura, Japan; and Edwin Baker, age thirty-eight, on 10 October 1891 married Oka Tori, aged twenty-eight, born at Tōkyō, both now residing in Himeji.[23] One measure of the extent of these relationships is offered by the *Kōbe Chronicle*, which in 1898, on the eve of the handover of the Concession to the Japanese, reported that the one elementary school at Kōbe served 20 Eurasians made up of 7 boys and 13 girls. Since their Eurasian designation was invariably defined by the Asian race of the mother, the *Kōbe Chronicle* enumerated fathers as follows: 2 English, 4 German, 4 French, and 1 Chinese.[24]

The case of the Kirby brothers provides an insight into this pattern. E.C. Kirby's obituary makes no reference to a family, yet he had a Chinese wife, born in Hong Kong, and with her he had three daughters. Although his will indicates that she was his housekeeper, there is evidence that she was also his legal wife. Such relationships were not uncommon in the community. E.C.'s nephew, Alfred Kirby, had a common law relationship with a Japanese woman, said to be an inmate of the Ōtsu Licensed Quarter, whom he met while working on Kirby-built steamships on Lake Biwa.[25] Together they had two daughters, Mary (also known as Ume), and Amy. Evidently following E.C. Kirby's suicide, his wife and daughters spent the remainder of their lives in Exeter, England; at least one daughter married a British man. In other cases, widowed consorts with children were not so fortunate and had little choice but to try to fit themselves back into Japanese society either by having their children adopted by their own family or by forming new partnerships with other men.[26]

There were of course conventional marriages between men in the port and women drawn from their own backgrounds, although not always from their home country. The U.S. consular record provides a surprisingly busy round of such weddings in the port. For example, on 16 December 1871, William Horace Morse, age thirty-one, an employee of Smith, Baker & Co. and formerly of Boston, was married to Sarah Virginia Center, age twenty-nine, of Herkimer, New York, by the Reverend Arthur Morris, an Episcopalian clergyman from Ōsaka. In this case the bride's family were also engaged in the Japan trade; the Center firm appears in the record of shipping as the agents for the Pacific Mail Steamship Company.

The Presence of Missionaries

American missionaries formed one of the most identifiable and coherent strands of the American population in the port. While British and French clerics appeared on the scene and established churches and schools in Kōbe, American religious zeal was disproportionately more vigorous, reflecting the very strong evangelical foundation of American Protestantism in the second half of the nineteenth century. The practice of Christianity as well as all forms of evangelism had, of course, been prohibited by Japanese authorities following the expulsion of the Jesuits from Japan in the seventeenth century. That prohibition quickly became a source of tension when Japan re-engaged with the West following Commodore Perry's intrusion into Tōkyō Bay in 1853 and 1854. While the Japanese ultimately acceded to the foreigners' demands that clergy be permitted to reside and practise religion in the treaty ports, missionaries were initially forbidden by the Japanese authorities to proselytize outside these restricted areas. Soon, however, many religious groups began breaking the nominal prohibition on travel outside the Concession, either by exploiting their ability to teach Western ways and language, including religion, to an eager Japanese population, or by discretely disappearing into the countryside to conduct their evangelism. The oft-cited role that the bicycle played in these activities provides a vivid reminder of technology's role in missionary work; it offered mobility and at the same time drew the curiosity of the Japanese whom the missionaries were hoping to convert.[27]

The first Christian religious service in Kōbe was conducted by the Reverend Father Pierre Mounciou, a French priest of the *Société des Missions Étrangères* of Paris, in July 1868. A weekly mass was held at a makeshift chapel in Moto-machi in the Native Town near the Belle Vue Hotel. Shortly afterwards, a Roman Catholic church was built on lot #37; it was consecrated in April 1870.[28] While nominally this was a French Roman Catholic initiative, it probably served a broader Roman Catholic constituency within the foreign population of the port throughout the treaty port period. By 1877, members of the Sisters of the Infant Jesus of Chauffailles were present in Kōbe, where they opened an orphanage and private school.

Protestant services were slower to begin. It is likely that occasional services were conducted by Church of England clergy passing through the port in 1868–9.[29] Episcopalian missionaries of the American Church Mission Society, led by Channing Williams, were active in Ōsaka when it opened; the group's first recorded presence in Kōbe was in 1875, when Charles Frederick Warren and his wife, along with Henry Evington, appeared there and built a small chapel using funds collected locally.[30] Surprisingly, although the British were the dominant national group in Kōbe from the opening of the settlement, the Church of England was a latecomer there. It appears that British efforts through the Church Mission Society were initially directed to Ōsaka; services were not begun in Kōbe until 1876, under the auspices of the Reverends Hugh James Foss and F.B. Plummer. The cornerstone for St Michael's Church was not laid until 1881, by Foss, on behalf of the Society for the Propagation of the Gospel (SPG). Foss served Kōbe's foreign community for several more years after that, also overseeing an evangelical mission to the Japanese conducted by other members of the SPG.[31]

The strongest Protestant missionary presence, however, proved to be the American one. The Reverend Daniel Crosby Greene of the American Missionary Board arrived as early as 1870 and sought to commence a regular Protestant service. His early initiatives were anything but a roaring success.[32] His first services were conducted in the "hot and stuffy" Masonic Lodge, newly erected on lot #18. In November 1870 and in the months following, a series of meetings were held in the hope of energizing an initiative for a Protestant Chapel. Attendance at these meetings was remarkably sparse, consisting often of a handful of consuls, who presumably had been invited specifically to offer leadership.

Finally, in April 1871, a large enough group was assembled to commit to the project. Perhaps more important, the owner of lot #48, Mr Bradfield, offered to give half the lot for the raising of a church provided that the land be used for this purpose "forever."[33] Until funds were secured for a church, Greene held services in his own home.[34] Not until 1874 was a church building erected on this site, and it soon became the centre for the settlement's growing band of American missionaries.

The American missionary presence at Kōbe was the child of the American Board of Commissioners for Foreign Missions and the Women's Board of Mission of the Interior, an arm of the Congregationalist denomination and therefore strongly rooted in New England. Initially the sole spiritual leaders were the Reverend Greene and his wife Mary Jane Forbes Greene, a formidable missionary presence in her own right. Before long, others arrived, until by 1875 the American board could boast thirty members operating in various locations in Japan. As part of this effort, the Greenes moved to Kyōto; they were succeeded at Kōbe by the Reverend Orramel H. Gulick, most likely in the company of other members of his extended family.[35] The Gulicks were part of a New England missionary clan who had conducted missions in Hawaii before moving on to Japan. In mid-December 1875, the American Consul recorded the untimely death of the wife and son of one member of this clan; two years later, he also recorded the death of Orramel's father, the Reverend Peter Gulick.[36] The intrepid traveller Isabella Bird, who had known the Gulicks from her stay in the Sandwich Islands (Hawaii), travelled to Kōbe soon after, specifically to visit the Gulicks, and her book *Unbeaten Tracks in Japan* offers a sympathetic and revealing glimpse of their activities at Kōbe:

> This is the other headquarters of mission-work under the auspices of the "American Board." Somehow when one thinks of Kōbe it is less as a Treaty Port than as a Mission centre. It is partly to see the process of missionary work that I came. Everything is at high pressure, and a hearty, hopeful spirit prevails among all who have got over the initial difficulties of the language, which press heavily on newcomers. The missionaries are all intensely American in speech, manner, and tone, and set about their work with a curious practicality and a confident apportionment of means to an end which I have not seen before in this connection. They are quite a community, mixing little, if at all, with the other foreign residents, but forming a very affectionate and intimate family among themselves. Kōbe being a place of energetic effort, and of reputed success, is the spot in Japan in

which to gauge in some degree the prospects of Christianity; but I shall defer saying much on the subject till I have been in Kioto and Ōsaka.

... Yesterday evening we went to tea at the "Girls' Home," (p. 223) a boarding school for twenty-seven Japanese girls, the prettiest house in Kōbe, in very attractive grounds. This is conducted by three ladies, with Japanese assistance. The girls live Japanese fashion, but learn our music, in which they are very anxious to excel ... The Mission has at Kōbe nine men missionaries, all but one with wives, and five single ladies: in Ōsaka four men and three single ladies, and in Kioto, three men and one single lady. Two are medical missionaries, and through their popular work several villages within the treaty limits have been opened to Christianity. In Kōbe, as elsewhere, there is a complete separation between the foreign and missionary community ... most foreigners speak of them as a pariah caste.[37]

Further evidence of this missionary enclave is revealed by the U.S. Consular Register, which documents the personnel associated with this missionary society.[38] Between 1880 and 1884 there was a good deal of coming and going with respect to the personnel forming this group. In all, the register records twenty-one individuals, including the wives and children of missionaries; as Isabella Bird indicates, the active annual adult cohort was about nine. At the core of this group during this period were the Reverends Henry Davis, W.W. Curtis, DeWitt Jenks and his missionary wife Sarah, and Dr John Berry, a medical missionary. But mention must also be made of the important presence of single female missionaries, such as Eliza Talcott and Julia Dudley, two members of this band sent by the Women's Board, who together founded a school for Japanese girls in 1875 under the patronage of the Viscount Takayoshi Kuki. This institution evolved over the years by adding college courses; ultimately, in 1894, it become Kōbe College, which along with sister entities in Tōkyō and other centres in Japan became one of the early Western-inspired institutions of higher learning aimed at Japanese women.[39]

As Bird notes, the missionaries' contacts with other members of the foreign community were few. It is important to appreciate the zeal and drive of this missionary group. Recent studies, particularly those focused on female missions, have emphasized the emergence of practical and social-activist missionary activities in America. This impetus grew out of the second Great Awakening and gained strength as missionaries from the northeastern United States poured toward the Western frontier and commenced work with native groups on the Great

Plains. Sustained by the pieties of this religious tradition, this group became willing agents of the American doctrine of Manifest Destiny. Missionaries were soon moving into the Pacific, bent on bringing spiritual and social enlightenment to what they regarded as the heathens of East Asia.[40] The women missionaries were confident in and committed to their work, which benefited not only from their experiences as battlefield nurses during the Civil War, but also from their success at cracking open the formerly male-dominated colleges and universities. Not surprisingly, those who found their way to Japan in the 1870s were what one author described as "feminists under the skin," and they can be linked directly to a societal transformation that resulted in the Temperance Movement, the Suffragette Movement, and the entry of women into a broadening range of professions. It is not surprising that missionaries in general, and these highly active female missionaries in particular, would be viewed as antithetical to the social mores and rhythms of the mercantile class that surrounded them in treaty ports such as Kōbe.

The drive to found missions in Japan, using the treaty ports as bases, was maintained throughout the period. Another American missionary worth noting was the Reverend James William Lambuth of Alabama, a member of the Methodist Episcopal South denomination. Lambuth had been active in China during the 1850s and arrived in Kōbe about 1886 in the company of his son, the Reverend Walter Russell Lambuth. After settling in the Concession, the Lambuths began establishing schools to serve both male and female student converts among the budding Japanese Christian community around Kōbe. The elder Lambuth died at Kōbe in April 1892, but his son carried on the work. By the time he returned to the United States around 1910, he had established Kwansei Gakuin, an academy that prepared Japanese men for the ministry, Seiwa Jo-gakuen, a school for women, and the Palmore Institute, a night school for men and later for women.[41]

The Social Fractions of the society in Kōbe

Who were the people who arrived to breathe life into this new and far-flung outpost of the Euro-American world? To the extent that biographical details can be pieced together, it appears that most of the newcomers in 1868 were anything but new to East Asian commerce and port life. Most arrived in Kōbe after sojourns in the previously opened Japanese treaty ports of Yokohama and Nagasaki, or from Shanghai or Hong Kong. They were the inheritors of a culture built around the China trade, which itself was closely affiliated with the British-Indian culture of the day. These were men who had a firm sense of their own superiority and civility – an attitude that stemmed more from the refinements and lifestyle they affected than from their commercial acumen. Indeed, many in this crowd had achieved their place in the East Asian trading world as "opium smugglers and ruffians."

But it can also be said that the 1860s were marked by the convergence of a number of factors that encouraged the Far Eastern trade and the participation in it of men of small capital. The development of the Suez Canal, and the opening of regular ocean shipping using steamships, made it possible for men of limited resources to compete with larger firms, particularly in the era before the telegraph linked the Far East to European and American markets. This trade was, as the British Consul at Yokohama pointed out, a "game of chance," for the interval between dispatching a cargo of trade goods from Japan to Europe and learning whether a profit had been made could be as long as nine months.[42] The 1860s were also the years of the American Civil War, a calamity that resulted in many citizens of the vanquished Confederacy becoming economic refugees.

Table 7.3 provides one window onto the occupational structure of the foreign community soon after its opening. It is always difficult to project from occupation to a social-class attribution from the vantage point of a more modern age, for the implied social meanings of professions and job categories have changed in the intervening century. Nor can we assume that these occupations had the same nuanced meanings across the several nationalities represented in the table. Nevertheless, an attempt has been made here to group these occupational or business entities into functionally connected categories as one means to examine the structure of this population economically and socially. What is apparent from this evidence is that apart from the "Religious" category, there was a strong functional symbiosis among the occupations found in the port. This community was sharply focused on mercantile trade, and while the numbers of merchant traders were limited, they were supplemented by a multitude of essential services that supported trading activity (e.g., the nautical trades and professions) and that provided services necessary for maintaining a Western lifestyle. But sorting this picture out into a more textured social fabric presents an interesting challenge.

Table 7.3. Businesses and occupations recorded for foreigners at Hiōgo, 1871

	Br	US	Ger	Dutch	Fr	Aust	Swiss	Other	Total
Mercantile									
Merchant firms	9	4	7	5	3		1		29
Banks	2								2
Brokers	6				1				7
Professional									
Civil engineers	1				1				2
Architects	1			1					2
Chemists								2	2
Doctors	1	1		1					3
Services									
Auctioneers	4		1						5
Newspapers	2								2
Stores	5	3	1			5			14
Hotels	1			1	2				4
Billiard parlours	1				2				3
Mail agents		1							1
Tailors	1		1						2
Watchmakers			1		1				2
Livery stables	1	1							2
Hairdressers					1				1
Grog shops	1	3			1				5
Nautical Trades									
Ships' carpenters	1	1		1					3
Ironworks	2								2
Ships' *compradors*	3	1	1						5
Pilots	1	2							3
Religious									
Missions		1			1				2
Churches					1				1
Total	43	17	13	9	14	5	1	2	104

Source: FO 262 #75. British Foreign Office, Letters to Consular Office in Kōbe and Yokohama.

Probably three distinct groups landed up on Kōbe in the early stage of its existence. One group were the sailors, a transient population who drifted back and forth along the trade routes forming the broader Asian and (to some extent) Australian network of commerce. The advent of regularly scheduled steamship routes was beginning to transform maritime life for these men. However, a good deal of trade still involved sailing vessels with ragtag mixed crews hired on in each port at the outset of a voyage. Thus one of the first foreign presences in Kōbe was undoubtedly the "sailor town," which was absorbed into the Native Town west of the Concession. Surprisingly little is recorded of this zone and its occupants, except to note that the Foreign Concession was relieved of housing this element and the grog shops and bordellos that accompanied it. In a significant sense, the absence of this zone, or its hidden nature, caused many visitors to Kōbe to remark on the comparative gentility of the town compared to Yokohama, which was known for its steamy "swamp town" and its licensed pleasure quarter.

The second group forming the society of early Kōbe were the owners of small shops, hotels, and nautical repair yards. Most of the latter came from Britain, the United States, or the Netherlands, but beyond that we know very little, since they, like the sailors, tended to locate on the edges of the Concession. For example, by the end of 1869, the "Hiōgo and Ōsaka Foundry and Ship Yard located at 82 Shindin, Western Beach, Kōbe advised that J. Wignell Engineer will do all engineering, boiler, blacksmith work, will furnish castings in brass and iron to order, etc."[43] The service providers in the Concession and the Native Town served an important function for the trading community; as provisioners, tailors, restaurateurs, and hotel and livery keepers, it was they as much as anyone who helped sustain the port's Euro-American lifestyle. Some in this group were women, whose identity we glean from business advertisements in the local papers, and their presence helped balance what was a strongly male society. An example was Mrs Bisson, the manager of the Belle Vue Hotel on the Native Bund in 1869.[44]

It is difficult to get a clear fix on who these people were and where they came from. It seems very likely that most had been in East Asia for some time, moving from port to port, trying their luck in small service trades and businesses. For example, the auctioneer and newspaper publisher James Wainwright, who unfortunately lost his life when the steamship *Haya Maru* sank in June 1869, had come to Kōbe from Shanghai, where he had been since 1862.[45] Wainwright was an American, born on the eastern shore of Maryland, who had, after stops in Philadelphia and Pittsburgh, followed the Gold Rush of 1849 to San Francisco, where he raised a family. As a Confederate sympathizer, he found himself at odds with associates in San Francisco, so he sought new horizons in China. His lifelong wanderlust ultimately brought him to the newly opened port of Kōbe in 1868, where he was one of the first to come ashore.

There appears to have been a fluidity to the peregrinations of men like Wainwright, suggesting that many in this category began with limited capital and that each move they made was either a flight from debt or an attempt to cash in on brief success before striking out to the next new commercial frontier. One place that seemed to have contributed elements to Kobe's new population was Nagasaki. There had been a European presence there longer than anywhere else in Japan; as we have seen, the port had been the sole location where the Bakufu had tolerated Westerners. So it was natural for Nagasaki to be one of the first ports to open under the treaty arrangements. It also enjoyed the logistical advantage of being close to Western firms already established at Shanghai. Furthermore, Nagasaki was proximate to the progressive regional forces associated with the Satsuma and Chōsū *hans*, both of which were aggressively acquiring Western ships and armaments. However, after an initial burst of trade at this location, it became clear that Japan's economic centre of gravity lay to the east, which consigned Nagasaki increasingly to the periphery. Several trading firms moved their agencies from Nagasaki to Kōbe when it finally opened, and it seems likely that many of the supporting community relocated as well.

While members of this "class" of petit bourgeois rarely grew wealthy by their labours in the port, a few did, and their cases bear noting insofar as they illustrate the rigid social stratification that marked this community. The case of Edward C. Kirby and his brother Alfred, the principals of Kirby & Co. and Kōbe's most significant foreign industrial entrepreneurs, provides an insight into this pattern. One of the early entrants to Kōbe, the firm was initially engaged in shipbuilding and repair.[46] Its founder was Edward Charles Kirby. From his obituary we learn that he was born in Stourbridge, Worcestershire, where as a young man he was apprenticed to a druggist.[47] Evidently unhappy in this pursuit, he found his way to the Australian colonies as part of the tide of gold seekers in 1855 or 1856. In 1860 he travelled to Shanghai, where he again tried several trades before moving on to Ningpo, where he opened a store and chandlery as an associated branch of a Shanghai firm. In due course he also opened a hotel.

When the rebellion in Japan ended in 1865, he moved to Yokohama, bringing with him a shipload of goods, which he used to start a business supplying English and French naval fleets. His fledgling firm specialized in baking, butchery, and general stores, often employing twenty Europeans and a large staff of Chinese and native Japanese. He evidently introduced the first bread machine to Japan, using supplies of flour obtained from San Francisco. At the opening of Kōbe in 1868, he established a branch of his business there, all the while continuing to pursue opportunities in Yokohama, where he began to construct speculative buildings for the foreign business community. As trade in Kōbe increased, he committed himself to the port, selling his extensive businesses and contracts in Yokohama to Messrs Lane, Crawford & Co. of Hong Kong.

In 1872 he began constructing a brick business block on lots #13 and #14 in Kōbe. However, as early as 1869 he had surmised that Kōbe must become a place of shipbuilding and engineering, so soon afterwards he started a modest shipyard on a beach lot at Ono. In 1873 he leased a lot about a quarter of a mile from the Foreign Concession east of the Ikuta River; this became the Kōbe Iron Works. This location had the benefit of water frontage, and the firm built a wharf and sheer legs for lifting heavy weights. Vessels drawing 18 feet of water were able to come alongside the wharf. This provided a distinct advantage, since at the time there were few other deepwater docks in the port and most cargo had to be transferred to and from shore by lighter.

Kōbe Iron Works was a partnership with Kirby as the principal shareholder, alongside Robert Huggan, one-sixth share, and John Taylor, one-sixth share. Together they formed a partnership, named the Kōbe Iron Works Company, as engineers, iron founders, blacksmiths, and shipwrights for seven and a half years with capital of $20,000. In 1877 Huggan sold his shares to E.C. Kirby, and in 1879 Taylor did the same. At this point Kirby expanded the yard. He also persuaded his brother Alfred, who had been working in Karachi, to join the firm as Director of Engineering. The record shows that in addition to providing salvage and repair services, the firm built and in some cases operated a number of small steam vessels. E.C. Kirby took his own life in December 1883, an act apparently precipitated by the foreclosure of his firm by the Hong Kong and Shanghai Banking Corporation, from which he had borrowed 50,000 Mexican dollars to expand the yard in 1879. While there is no evidence to confirm it, there are hints that Kirby had run afoul of the Japanese government, which was attempting to wrest control of shipbuilding. One writer believes that Kirby's finances were sound and that the squeeze placed on him by the bank was unjustified.[48] Whatever the case, the firm was transferred to the Japanese authorities, whereupon it became known as the Onohama Naval Dockyard.

Kirby's life in Kōbe offers us a glimpse of the arcane social stratification that existed among the port's expatriate community. As someone not engaged in mercantile trade, Kirby was deemed ineligible for membership in the local clubs. This was a source of some bitterness for Alfred Kirby, who contended – probably on solid grounds – that E.C Kirby was superior in education, family, and wealth to the bulk of the club members.

Finally, there were the merchants – the traders and their agents, junior clerks, and ship's masters. It is perhaps misleading to lump these men into a cohesive class, especially given the strong class consciousness that marked most Europeans in this era. Included in this group, although at times they were perhaps self-consciously socially distant, were the foreign consuls. Only the British, American, and French governments regularly sent their own staff to Kōbe. Also, most American representatives were political appointees who were being rewarded with a consulship for service in the military or some other realm of public life.[49] Many of the British consular staff were themselves old hands in the Far East, moving from port to port as assigned and only rarely returning to their home country. Most other foreign governments appointed a resident merchant to serve as consul, hence social distinctions were rarely as sharp as implied.

Another strand of this class were men attached to established trading firms, who arrived in Kōbe with a decade or more of experience in the Far East, including Japan. For example, Kenneth Ross Mackenzie had been working in Shanghai for Jardine Matheson & Co. when in 1859 that firm sent him to Nagasaki to establish a base there.[50] When Kōbe opened nine years later, he was again among the first on the ground in the service of his firm. By the time Kōbe opened, many of the larger trading firms were well-established multi-agency enterprises. These firms all had different national origins and histories. Some were well capitalized and multifaceted; others were small, consisting of commission merchants whose operations were cobbled together opportunistically and often unravelled as quickly. Nevertheless, it is worth looking closer at some of these firms to capture their characteristic patterns and social origins.

One firm that succeeded early on at Kōbe was Walsh & Co., an American firm, later known as Walsh Hall & Co. In the port's first years, this firm was the pre-eminent America *hong* in Hiōgo, occupying lot #2 on the Bund, on which it built what was perhaps the most impressive headquarters of all those fronting the harbour. The firm had been founded by the four Walsh brothers.[51] Sons of an Irish immigrant who settled in Yonkers, New York, the four boys somehow got to Japan about 1855, apparently after working their way to China with stops in Africa and India. They worked in the China trade in Shanghai and other ports. The circumstances of their arrival in Japan are unclear. It is believed that Richard, the eldest, did not stay, but it appears that his wife and her brother were in Japan in 1874 or 1875 and that she died there in 1876. The children of this marriage were sent back to the United States to be raised by unmarried sisters in Yonkers and Wayne, Pennsylvania. It is known that John Greer Walsh was appointed Acting U.S. Consul at Nagasaki by Townsend Harris in 1859, a position he evidently held until 1865. Apparently this was an unpaid position, for Walsh carried on business as a means to live. In concert with his brother Thomas, he organized a firm initially called Walsh & Co. The firm conducted business in both Nagasaki and Yokohama and later, in 1868, opened a branch in Kōbe under the management of Arthur O. Gay. Later, John Greer Walsh relocated to Kōbe from Nagasaki. It appears that, like other traders, he married a Japanese woman, Rin Yamaguchi. That was around 1862, and he had a daughter with her, Aiko Yamaguchi (1864–1910).[52]

The firm operated in Kōbe's Native Town until the Concession had been laid out and land was available for sale. Its first premises were in a building opposite a small landing stage, or *hatoba*, for sampans, and because it flew the American flag from a flagstaff, the landing stage became known as the Merikan Hatoba. In the first land sale the firm purchased lot #2, and in due course it built magnificent business premises, later described as an immense, two-storey, wood-frame building faced with cement (see Figure 6.6, p. 134). It contained "extensive offices on the ground floor and palatial residential quarters above for the manager, with drawing room and dining-rooms grand enough for a mayoral reception."[53] This building was decorated with ornamental "demon tiles" along the gable roof. As Williams notes, unlike the customary tiles on such buildings, which employed either an image of the company trademark or Chinese characters that served as a charm to ward off bad luck or fire, these tiles rendered the phrase "America Number 2" in Chinese characters. In this way, the Walshes were declaring their nationality and asserting their pre-eminent place in the port's trading landscape.

The firm specialized in the export of silk and tea. Later it took on partners Frank and George Hall and became the leading rival of Jardine

Matheson in the port. The failure of the silk trade at Kōbe led its merchants to explore other exports. One of these was the shipping of Japanese cotton rags to paper makers abroad. These rags were plentiful in Japan, and exporters competed with Japanese dyers, who sought rags to extract indigo dye from them.[54] Evidently these rags contained a lot of lime and moisture, and without realizing this, Walsh & Co. packed the rags tightly into ships, only to find that spontaneous combustion resulted, with heavy losses. As a result of this experience, they began to consider making paper within Japan itself. Paper making using Western techniques and imported machinery had begun in 1874 in Tōkyō with the opening of a plant there; another company followed soon after that would ultimately become the Oji Paper Co. The Mita Paper Co. was started about the same time. In Ōsaka the government established a plant in 1875, and the Umezu Co. (a.k.a. Papier Fabrik) was founded in Kyōto as a prefectural mill, with German assistance.

In 1875 the Walsh brothers decided to open a paper mill, and sent their younger brother Robert to procure heavy paper-making machinery in the United States with a view to starting operations the following year.[55] In the meantime, a company known as the Japan Paper Co. was floated, with the Walsh brothers and their English partners each owning half, and a site was sought for a factory in either Yokohama or Kōbe. In the end, a site in Kōbe was chosen, just beyond the upper margin of the Concession on farmland at Sannomiya. However, this attempt seems to have produced unsatisfactory results from a business point of view, and the company soon had to be reorganized under the direction of Walsh Hall & Co. Its name was changed to Kōbe Paper Mill in 1877.

It has been claimed that the company played a leading role in developing industry standards and in asserting the interests of private entrepreneurs in the face of government-supported paper-making competition. One partner in the Kōbe enterprise was Yatarō Iwasaki, the principal business mind and driving force in the Mitsubishi firm. The Walsh brothers had fostered this connection earlier, and they apparently provided Iwasaki with capital to purchase ships and merchandise. Having profited from these arrangements, Iwasaki offered interest-free loans to the Walshes when they began the paper company.[56] After J.G. Walsh died, the entire firm eventually moved into the hands of the Mitsubishi Company, although the firm's name was not changed until 1904, such was its esteemed reputation.

At the other end of the mercantile and business scale were the commission merchants or agencies. One of these small-scale firms was

Browne & Co., founded in 1869 and consisting of Henry St John Browne and L.R. Goldsmith. In its early days, the firm organized imports and exports and served as the representative for a number of insurance companies, including London and Oriental Steam Transit Insurance, British and Foreign Marine Insurance, and the Union Insurance Co. Later, it would act as the agent for a number of steamship lines. When the tea trade rose to dominance in Kōbe, Browne & Co. was one of the smaller houses involved in it.[57] Browne, who would play a prominent and ongoing role as an elected member of the Municipal Council, also invested in the Kōbe Ice Co. and the Hiōgo Gas Co., which had the contract for street lighting in the Concession.[58] Through these entrepreneurial activities, the firm grew to the point that by 1877 it listed among its employees Goldsmith, M.T.B. Macpherson, and C.D. Bernard.[59]

Where did the Chinese fit into the trading community of Kōbe? In sheer numbers, they were at least half the port's foreign population, and they had arrived early, with the first of the Western traders. They performed many of the basic functions of trade as labourers in the godowns and tea firing rooms, but perhaps even more critically *shroffs*, *bantos*, and *compradors* in the trading firms' offices. In these positions they handled trade transactions, conducted communications with the Japanese, exchanged money, and provided domestic services within the traders' living quarters. Undoubtedly many of the Chinese had acquired these skills in the China treaty ports, which is where this symbiosis between Western traders and local Asians had developed. In time, some Chinese became traders in their own right, even though this first required that they depend on others to conduct their affairs with respect to the customs house.[60]

By the 1880s, Chinese merchants were conducting an important trade in marine products, such as seaweed and dried fish, all of which supplied Asian markets in Japan and China. Socially there remained a wide gulf between these two components of the trading culture. Some of the Chinese staff, particularly those performing domestic duties, may have lived in-house, but most of them resided in the Native Town, where they eventually formed a Chinatown.[61] Clearly, then, the missionaries were not the only group who did not fit into the social mainstream as it took shape in Kōbe. From the beginning, the Chinese were a persistent and self-conscious presence there, one that has endured to the present day.

Chapter Eight

Measuring Success in the Concession

For all the bravado and self-important assumptions and affirmations of those who formed the trading culture at Kōbe, and for all the accolades accorded to the Concession as an ordered and well-run multicultural outpost of the Euro-American world, just what was life really like there before the treaties ended in 1899? Were those who occupied this curiously artificial place exemplars of modernity, or were they unwitting performers in a pastiche of the Victorian world? Were they a kind of zoological garden whose actions and behaviour were selectively read and absorbed by their Japanese hosts?

Social Life among the Foreigners

What was day-to-day life like for foreigners residing in the port? First, it must be understood that the seasonality of much of the trade underpinning the place was such that life in the mercantile community involved periods of intense business activity punctuated by slack periods. Moreover, as noted earlier, the treaties specified that foreigners could not travel into the port's hinterland beyond a geographic perimeter of 40 kilometres from the Concession without an internal passport from the Japanese authorities. This meant, for example, that they could travel to Ōsaka but not to Kyōto. And even if a passport were obtained, travel was complicated both by the foreigner's lack of language skill and local *savoir faire*, and by the wariness of the Japanese, who were equally unfamiliar with foreigners. These impediments would break down as the period progressed; even so, the reality was that the foreign residents had to make their life within a very narrow set of social spaces. Not surprisingly, a peculiarly self-conscious set of social rituals and activities developed that served the dual purpose of greasing the wheels of commerce and relieving the boredom that was the fate of those living in Kōbe. Because this was largely a male society, the resulting activities and lifestyles catered to men.

Social Clubs and Sporting Activities

We can assume that the first two years following the opening of the port would have been consumed with acquiring and building housing, developing intelligence on local commodity supplies and trading partners, and forming the basic infrastructure for trade. Yet there is evidence that very quickly these men also set about developing leisure activities. The hotels that sprang up played a significant role in this. For example, what would later be called the Kōbe Club had its genesis in April 1869, when twenty-seven gentlemen of the port began meeting at the hotel operated by Mr Dutronquoy in the Native Town. The club provided a reading room, a billiard parlour, bowling facilities, and of course a bar – all of the necessary accoutrements of male sociability during that era. Very quickly the club went through a sequence of names – the Hiōgo and Ōsaka Club, the Union Club, the International Club – shifting locations with equal ease. By 1870 the club was operating out of the Oriental Hotel on Kyo-machi. By 1876, it had settled on the name the Kōbe Club.[1]

More or less concurrently, the German merchants in Kōbe started a club that came to be known as the Concordia Club and that for some time welcomed non-Germans into its fold. This club, initially located just outside the Concession on the eastern flank, provided similar recreational attractions as well as a large upper-level veranda. Later it would build a substantial club premises on Kyo-machi adjacent to the Oriental Hotel. However, the Franco-Prussian War (1871–2) created tensions among the international members of this club; membership declined, and the club found itself stretched to maintain its facilities. At this point it sold its property to the rival International Club and took over the latter's rented premises in the Oriental Hotel.[2]

Membership in these organizations seems to have been restricted to those who were involved in import–export commerce, and one might speculate that the club provided a means for these men to exchange intelligence and pool resources, very much as the traditional mercantile coffee houses had in an earlier age. It is difficult to know how membership expanded or waned, given the fluidity of the settlement's population caused in part by swings in the local economy, such as the downswing of the 1870s. Nevertheless, by 1888 the *Hiōgo News* was able to report that the Kōbe Club numbered 132 members. Moreover, "the well-situated lot 74 [located on the northeast corner of Kyo-machi] and the fine block of buildings on it have just been acquired [from] the Club Concordia for $18,750." The editor noted that "this is a rising club, whose members pledged $20,000 for the purpose."[3] Finally, in 1890, the Kōbe Club relocated to a grander new brick building on the Hill in Kanocho.[4] By then it had become the premier gentlemen's club in the port and maintained a strict dress code. Women, of course, were not admitted under any circumstances.

In the meantime, other organizations had sprung up to support those residing in the community. Among these were two Masonic Lodge chapters. The first of these was the Hiōgo and Ōsaka No. 498, chartered under the Scottish Constitution, which commenced in 1870; the other, organized under the English Constitution and styled as the Rising Sun Lodge, dates to November 1872. Very soon after their formation, a Masonic Hall was constructed on Kyo-machi, sandwiched between the Oriental Hotel and the Kōbe Club. This facility served both lodges and was also a location for balls and theatrical performances.[5]

Sporting pursuits were central to middle-class male life of the time, and those entering the ports seem to have found outlets for these vigorous pursuits initially by staging games, regattas, and horse races.[6] In time, cricket and football teams were established, which offered the possibility of matches (and rivalries) between ports.[7] These and other team sports required dedicated facilities and investments. When the Japanese authorities relocated the channel of the Ikuta River in 1871, it became possible to transform the former river bed into a recreation ground, under the management of the Municipal Council. By what seems to have been a mutually supportive agreement, the leadership for making this space conform to the expectations of the residents fell to the Kōbe Regatta and Athletic Club (KRAC). The prime mover for this organization was Alexander C. Sim, a Scottish-born druggist, who was both a dedicated athlete and an energetic public figure in Kōbe, sometime member of the Municipal Council, and chief of the Concession's fire brigade throughout the treaty port era.

Apart from efforts to level the newly acquired ground and make it suitable for team sports such as cricket and football, KRAC built a gymnasium on the southeast corner of the Recreation Ground in 1878. In addition to its intended purpose, the gymnasium was used for theatrical performances and balls, as an alternative to the Masonic Hall. Maintenance of the Recreation Ground fell to the Municipal Council, and the council's minutes sometimes refer to it. For example, in February 1883 there was long discussion of the Recreation Ground's expenditures. The Works Committee led by A.C. Sims proposed to level new ground to match the existing cricket ground and lawn tennis courts. This would make the ground level to the northern boundary, but it would also require cutting down the "oak grove." Expenses were estimated at $1,444, a considerable sum to which two councillors objected. Both were German and perhaps therefore ambivalent about a cricket ground; that said, the basis of their concern seems to have been the call on financial resources that the council did not have owing to its inability to levy taxes.[8] On another occasion, the works superintendent, Herman Trotzig, reported to the council that the Recreation Ground had required a great deal of attention following a recent typhoon: the lawns were recovering, and the ladies' pavilion and watchman's lodge had been repaired. He further reported that the lawn tennis club had been allowed to lay some cinder courts on the upper ground.[9] In 1884, Trozig reported that Higashi-machi, the street separating the Concession from the Recreation Ground, had been reduced in width by 30 feet to enable an extension of the Recreation Ground. Council appropriated $800 for this purpose; the work was completed in June of that year, and a well was sunk to provide water for the trees and lawns. In addition, a path connecting the Recreation Ground to the Kōbe Club had been made by private subscription. The council laid a drain connecting to the one running along Higashi-machi and erected a new lamp at the Recreation Ground.[10] All of this suggests that the Recreation Ground featured significantly as a public resource in the minds of the Concession's leaders. Furthermore, this amenity did not escape the approval of visitors to the port. During his stay in Kōbe in 1881, Arthur Crowe was much impressed with KRAC and with the hijinks that revolved around it:

> There is a capital club – low rambling building equipped with every luxury, and situated within an extensive garden. Alongside lies the cricket-field and its adjoining lawn tennis courts. Within five minutes' walk of the cricket-field is the boating club, the gentlemen's rendezvous in the quiet summer evenings. The covered balcony overlooks the placid bay, in which various skiffs and four-oared boats dart about, contrasting strangely with the unwieldy, flat sampans. A pier for bathers, culminating in a spring-board, juts out some fifty yards from the beach. Never did we enjoy

such delightful bathing. In the evening the water was deliciously warm, and always clear, so that one could, without injury, remain in the water an almost unlimited time. The end of the pier is invariably the scene of numerous practical jokes, and probably is the merriest spot in (or out of) Kōbe during the evening of a hot summer's day. Should an unlucky junk be becalmed near the pier, a deputation of the best swimmers soon board the vessel (the crew being too frightened to resist), and cut adrift her sampan, which is towed in triumph to the pier by a tandem of mermen. Half full of water it is left near the beach to await arrival of one of the junk's crew, who, on the cessation of hostilities, swims ashore and punts back his boat.[11]

If vigorous manly pastimes served as a fillip to the boredom that punctuated economic life, the other outlet was the daily ritual of dining and drinking. Those belonging to one of the elite social clubs followed the ritual of breaking for "tiffin."[12] This routine might begin as early as 11:00 a.m., and never later than 12:30 p.m., and it led directly to the club bar. This pattern was repeated at 5:00 pm.[13] Murphy characterizes this behaviour as follows: "Rightful topics of conversation in the club were the interpretation of treaty clauses concerning imports and exports ... Good feeling, mutual charity and common ignorance triumphed." This was the "all-male environment of amiable sociability." With the help of alcohol, merchants "could escape from the pressures of ... their work ... find camaraderie with others who shared the same situation, call each other nicknames, and reinforce class, ethnic and perhaps most importantly, gender identities."[14]

In the club dining room and bar there was a fastidiously observed "pecking order," with the senior traders of the largest and most prestigious *hongs* claiming places farthest from the door and juniors spreading themselves out according to how long they had been resident in the port.[15]

The dining venue for others was one of the hotels or saloons in either the Concession or the Native Town. In these places, too, the rituals were adjusted by social rank. In addition to dining, devotees of this custom imbibed over lunch and most certainly in the afternoon and into the evening. This too invoked a certain ritual, which in the case of the Kōbe Club involved members shaking the dice as a means for determining who would pay for the round of drinks – a practice that inevitably required that members attend daily or risk the enmity of fellow members for "not holding up their end."[16]

Clearly club membership came with a built-in set of coercive practices, and the expansive consumption of food and alcohol placed a heavy demand on a man's purse. But here, too, the essential nature of this social behaviour offered a means to forestall the inevitable reckoning. A system of credit using "chits" allowed the member, in theory, to delay payment to the end of the month.[17] For those able to afford it, this system worked well and encouraged even freer indulgence. For those whose income was more limited, the credit system was really one of delayed reckoning leading to insolvency, even if the looseness of the accounting and procrastination might delay that fateful day for some time. In reality, most merchants were constantly in debt and were obliged to work out payment schemes that enabled them to stay afloat. Murphy suggests that most paid up within three years on average, but some expected that their obligations would only be cleared by their insurance policies at the time of their death.[18]

Club life and the rigid rituals of decorum in many ways mocked the boisterous and belligerent behaviour that many of these same men exhibited in other quarters. Among the lower orders, in the sailor's haunts, brawls and knife fights were all too common; these were expected behaviours of this class of men. But the men who were engaged in trade and other businesses could also display a darker and more violent side, and more often than not, the target of their attention was the Chinese or Japanese. Reports of Euro-American men abusing Asians abound and confirm the naked racism that permeated the thinking of those residing in the settlement. Some of this emanated from the cultural gulf and the foreigners' inability to communicate with the "other." The coolie labourers on whom the merchants depended were often subjected to hostility and abuse. It didn't help that these labourers, and those who pulled rickshaws, spent much of the summer nearly naked – cause enough to invite the condemnation of the settlement's European and American men and women, who were overdressed by social convention.

If sports and self-indulgent club and bar life, to which we might add the various options for carnal pleasures, offered recreational outlets to male residents of Kōbe, what was there for respectable women? Having been barred from the Kōbe Club, they had to make their own social life among themselves, resorting to endless rounds of "at homes," tea parties, and card games that were the standard for middle-class women of the day. The Concession provided few amenities specifically configured for women except for the landscaped promenade that graced the shore front along the Bund, where on summer evenings men and women might perambulate, to see and be seen. The occasional amateur theatrical or musical evening, or better yet, the occasional celebratory ball, offered another welcome break from the monotony of a desperately cramped and lonely social setting.

Part of the problem was that many of the women in the Concession were the wives of missionaries, a group regarded by the other women as stiff, judgmental, and humourless in their devotion to doing good works. It is difficult to tell to what extent these two classes of women formed a common rank. As noted, Kōbe did eventually establish its own churches, and these offered an accepted means for women in the community to participate in Sunday schools and Bible classes. But the real duties of women as wives and mothers would have revolved around managing the household staff and educating their children. Cut off by enormous distances from their homelands and the sympathy of family and friends, Kōbe's women seem to have been particularly vulnerable to their restricted circumstances.

The cramped confines of the port, the exceedingly narrow social sphere, and isolation from family and the familiar rhythms of life at home all exacted a toll on Kōbe's resident Westerners, both men and women. It was inevitable that crushing inertia infected them, as was captured by the *Japan Mail's* editor when he wearily and despondently wrote:

> We move in an eternal circle of similar dinners where a constant succession of similar talk interlarded with more than twice told jokes and stories blunt our intellect with wearisome repetition. In public and in private we see the same faces day after day until beauty itself loses somewhat of its charm and even official dignity declines toward the commonplaces of ordinary men ... So runs our life from year to year and every year is duller than the last, until at length we find ourselves metamorphosed into a callous race, almost denuded of nationality by the waves of eventless time and lulled into intellectual sleep by the absence of all healthy stimulus and mental exertion.[19]

What stands out in all of this is the absence of any apparent desire to connect with the Japanese in other than a utilitarian way. Given that the host Japanese society was making an enormous effort to catch up to the West economically and militarily, and was eagerly adopting many Western ways, the apparent absence of a meeting of cultures in places like Kōbe is surprising. This disconnection reinforces the impression that the foreigners were by their own devices and attitudes isolated and disengaged from those in whose midst they resided. In hindsight this cultural arrogance can be seen as a consequence of the zeitgeist, as emanating from a pervasive Euro-American cultural and economic imperialism as well as the social Darwinism that was an outgrowth of European colonialism.

The foreigners in Kōbe occupied a bubble of their own making. The Concession and its bucolic semi-rural appendages on the Hill offered an ordered and highly conventional landscape with most of the trappings of a middle-class town in the West. The calculated restrictions imposed by the Japanese narrowed the foreigners' movements; in those circumstances, the Europeans fashioned a way of life that did little to challenge their received cultural values. For most of the permanent residents of the Concession, that life was one of remarkable ease. Except for the palpable realization that they were half a world away from home, day-to-day life was perhaps little different than what they would have enjoyed in the provincial towns and rising suburbs of England, Germany, France, the Netherlands, or the United States.

Assessing the Economic Effectiveness of Kōbe

It is now generally accepted that the treaty port era produced mixed successes from the foreigners' point of view. The early "gold rush," which largely predated the opening of Kōbe, may have made some trading firms rich, but even this is now questioned.[20] After those early years, the trading *hongs* entered Japan with high expectations only to find the 1870s a "doubtful desert." As we have seen for Kōbe, an economy did develop based primarily on the export of tea and on the supply and servicing of the coastal trade in western Japan, including Kyōto and Ōsaka. The importing of cotton to be made into various textiles for export provided a balance to trade and a steady opportunity for Kōbe's foreign firms. Based on these staples, the port rose to rival Yokohama for trade supremacy. Yet as we have seen, the merchant community could hardly be called keenly involved in exploiting this opportunity with ingenuity, dedication, and purpose. Rather, the picture we have is of a community of sleepwalkers, who after establishing the apparatus of trade passed its operations to their juniors and Chinese employees. Preferring to indulge themselves in an artificial and (for many) agreeable lifestyle, they allowed events to overtake them as the Japanese mastered the levers of trade and modern manufacturing. The Japanese, chafing under what they regarded as unequal treaties, moved their country progressively forward until they again became masters of their own house.

Not without missteps, the Meiji era governments abolished Japan's feudal social order in favour of a more Western meritocracy, established a constitutional government that resembled those in the West, instituted a functional monetary system, began to create a modern military, built transportation infrastructure, and hastened to develop an urban system that ultimately allowed for a stronger national approach to education and to the fostering of social cohesion. By 1894 the Japanese had advanced to the point that the foreigners finally had to accede to their demands that the hated treaties be set aside. It was agreed that the treaties would end in 1899, at which time the affected ports would revert to full Japanese control. In the interim, relations between foreigners and Japanese slowly began to change.

As the day of handover approached, there were signs of strain, some of which underscored the tensions that surrounded a commercial system in transition. For example, the foreigners had long harboured suspicions about the honesty and integrity of Japan's customs administrators. Some of that suspicion was a consequence of poor cross-cultural communication and the Japanese fondness for arcane bureaucracy. In March 1898, however, the *Kōbe Chronicle* reported the arrest of fifty-five people for activities they had conducted through the Kōbe Customs House. Forty-two of them were committed to trial accused of fraud in the amount of 84,064 yen.[21] As the trial unfolded, it came to light that there had been a far-reaching conspiracy to defraud the Japanese government of duties on goods moving through the system. In 1896 and 1897, a number of people had been using forged seals and documents. Were these arrests an acknowledgment by the Japanese authorities that they needed to show they could manage their economic affairs at an international standard? Or was this a triumph for the foreigners, in that it highlighted the Japanese failure to conduct trade on terms that were transparent and above board?

In addition, tensions were rising between the foreign merchants on the one hand and the Japanese straw braid and cotton spinners' guilds on the other. The guilds eventually boycotted the firms of Samuel and Samuel, Finlay, Richardson & Co., and Fraser & Co.[22] Apparently, the Japanese had become emboldened in their dealings with the port's foreign merchants. Still other developments suggest that foreign traders were consolidating their presence in Kōbe as the date for ending the treaties approached. One group that stood out in this regard were the Chinese merchants of Ōsaka – sometimes referred to as the "Canton merchants" – who had chosen to relocate to Kōbe. Commentaries surrounding this move noted that as much as half of Ōsaka's trade was in the hands of these merchants and that Ōsaka had been significantly affected when, during the Sino-Japanese war, commerce there was brought to a standstill by the flight of the Chinese merchants.[23]

As the date of termination approached, practical concerns were raised regarding how to build a better climate for trade among the Japanese and foreigners. For example, the editor of the *Kōbe Chronicle* proposed that the local Japanese Chamber of Commerce and foreign Chamber of Commerce form a social union. He lamented the social gulf that existed between foreigners and natives after thirty years, noting that the foreigners would not have the right to vote in elections or to own landed property after 1899, but that they would in a sense become citizens of Kōbe.[24] Another expression of a growing accommodation was the recognition that Japan, after 1899, would still need foreign investment – most obviously, that it would need to convert its rail system to a common gauge.[25]

Exodus of Kōbe Residents

As the handover approached, the local newspaper recorded the departure of long-time residents. It is difficult to be certain of the circumstances of these departures. Some foreigners may simply have been rotating out of Japan, something that Europeans in port communities had always done. For example, when Mr and Mrs Von Krechi and Mrs H.E. Reynell and her children departed on the steamship *Hollenzollern*, it seems to have been for a furlough. H.E. Reynell had been in Kōbe at least as early as 1888 and seems to have taken over the well-established German firm of Kniffler & Co. But the tone of the reporting hints that for others, departure reflected coming uncertainties and the expectation that an era was ending. Mr and Mrs H.L. Baggally, who departed on the *Empress of Japan* after ten years in Japan, were described as leaving due to illness, and the editor lamented this loss to the community.[26] Others departing included Messrs H.C. Brushfield and A.L. Robinson, Mr and Mrs Delacamp, Mrs Ziegfeld, Mrs Doebbling, and seven children.[27]

There were other signs of preparations for the return of the Foreign Concession to the Japanese. In January 1898 the redoubtable Herman

Trotzig tendered his twenty-sixth annual report as the port's Works Superintendent.[28] In preparing to hand over the settlement to the Japanese, the Municipal Council has made extra-ordinary expenditures on the roads, main sewers, and side drains; all were said to be now in perfect order. To address flooding, the Japanese authorities had themselves begun enlarging the Division Street drain, intending to add an outlet drain to the main that would pass through the settlement. Fire wells were reported to be in good order. Trotzig reported that expenditures on the Bund lawn and on the Recreation Ground and Municipal Hall had been modest. Subscriptions to help pay for watering streets had been generous. There had been no serious thefts or crimes, and he commended the police force, which consisted of one foreign sergeant, two foreign constables, and thirteen Japanese constables.

Pay raises had been tendered for the latter and for municipal *bantos* and other employees. Finally – and this was no doubt a great relief to the foreign community in view of the impending changes – the Japanese authorities had granted a site for a new cemetery, to be located in Shioya, a comparatively new "suburb" on the Hill that would become a favoured residential location for the foreign community in the twentieth century. Later in 1898 the *Kōbe Chronicle* reported that Japanese firms were beginning to move into the settlement: the shipping firm NYK was opening an office there, and a tobacco firm from Kyōto was rumoured to be doing the same.[29] Stewardship of the settlement was transferred in July 1899.

Epilogue

After the Concession and the city of Kōbe were reunited administratively in 1899, the Foreign Concession retained its place as the centre of mercantile and financial activity. In the three decades of trade activity focused on this sector of the port, infrastructure and expertise had been developed that was too strong to abandon. Not surprisingly, the Concession would be the centre of gravity for post–treaty era business development. Because several key Japanese institutions, such as the *Kencho* (i.e., the Hyōgo prefectural offices), had been strategically placed close the boundary of the Concession, it was a simple matter to integrate these nearby Japanese institutions into the Concession, thereby making the latter the emerging city centre. Commercial life in Kōbe moved easily onto a new footing under Japanese sovereignty, and in this, the Concession provided a reinforcing model for urban development that resulted in broad streets and various civic amenities. These ideas and modes of design were progressively taken up by the Japanese, not only in Kōbe but also in Ōsaka and in other growth poles in Japan. The cadastral template created by Hart in 1867–8 and the transplanting of a quintessentially Western urban architecture would have a lasting impact on the Japanese. In the twentieth century this sector of the city inevitably was remade as Kōbe thrived. Gradually the merchants' *hongs* gave way to more modern business and financial edifices. As Kōbe grew outward and upward onto the Hill, a street railway system was built that reinforced the Concession's role as Kōbe's central business district. However, Allied bombing would exact a heavy toll in the late days of the Second World War; few of the Concession's original foreign buildings would survive it.

More than a century has passed since Kōbe's three-decade period as a treaty port ended. How are we to summarize this early phase of the city's development as a modern port? While Kōbe was among the last treaty ports to be opened, it made a stronger developmental thrust in early and middle Meiji Japan than many of the other treaty ports. Consider Nagasaki, a treaty port since 1859. As noted earlier, Nagasaki had been hosting a European enclave since the 1550s, which meant that it had enjoyed an initial head start over Kōbe. While Japan's long relationship with the West at Nagasaki was profoundly constraining from the perspective of the Westerners who were holed up on Dejima Island before 1859, Nagasaki had been able to absorb Western ideas and technologies, particularly during the late Tokugawa period. However, Nagasaki once it became a treaty port suffered the disadvantage of being far removed from the centre of action at a time when the Meiji Restoration was drawing Japan's creative forces toward the political and population centre of gravity that was the Yedo–Ōsaka axis. As Nagasaki's early advantage waned, its foreign settlement fell into disrepair as a consequence of losing its monopoly on Western contact. Kōbe benefited quickly from Nagasaki's collapse; many of the earliest residents of Hiōgo were merchants and tradesmen who had relocated to the newly opened port. Nagasaki's loss was Kōbe's gain.

Regarding the other treaty ports, Shimoda, Niigata, and Hakodate barely registered as Western commercial presences during the treaty port era. Shimoda, on the southern tip of the Izu Peninsula, 97 kilometres from Yedo, enjoyed an initial advantage as the first port to accept Commodore Perry's American expeditionary squadron. Under this pressure, it opened to American trade. But as the broader Western initiative to secure open ports unfolded, Shimoda fell away as a treaty port when Kanagawa – later renamed Yokohama – came into play in 1859. Yokohama's proximity to Yedo trumped any early advantage that Shimoda might have enjoyed. Niigata, on the Japan Sea northeast of Yedo, suffered from a shallow river harbour, and its opening to foreign traders was delayed until 1869. But even in 1874 the British Consul there reported that the harbour still required major alterations and that effectively there was no foreign trade in the port.[30] By 1880 the British authorities had ceased to include Niigata in their annual reporting. Hakodate, on the southern tip of the northern island of Hokkaidō, served as a base for foreign whaling fleets as well as commercial interests from eastern Russia. But it, too, struggled to find a sustaining economic rationale. The British Consul reported that even though Hokkaidō was heavily forested, construction timber had to be imported from elsewhere; the island's forests were exploited mainly for the manufacture of wooden matches, an industry that was centred on Ōsaka. By 1891 the port could report only four British and five Chinese trading firms and a total foreign population of seventy-three, including a single Russian. Thus it was a local hub at best, and its ability to grow was constrained by the lack of anything to trade with the West.

So the only two significant treaty ports were Kōbe and Yokohama. A comparison of these two ports as the treaty port era ended in 1899 provides a measure of the difference between the two. Yokohama had enjoyed an earlier start than Kōbe and had effectively cornered the market on the profitable silk trade. It enjoyed other advantages as well: with the advent of the Meiji government in 1866, imperial functions were shifted from Kyōto to Tōkyō, thereby consolidating the nation's capital there. As a consequence, western Japan fell into relative stagnation. But it is important to remember that Ōsaka, already a large city at the beginning of the Meiji period, continued to grow and expand its industrial base such that it had become by the end of the century the workshop of the nation, particularly as a result of its booming cotton industry. As such it concentrated much of the economic energy of the Kansai region,[31] and to some degree Kōbe became an economic adjunct to Ōsaka. As Ōsaka began to address the navigation barriers it faced, its own port began to challenge Kōbe. Thus the picture of Japan that emerges at the start of the twentieth century is of a nation with a pronounced core–periphery economic geography. Japan had two economic nodes: Tōkyō–Yokohama, and Ōsaka–Kōbe. Yokohama serving in effect as the port for Tōkyō, it came to enjoy a natural advantage as an export-led economy developed in the Kantō region. Because its foreign population was larger, Yokohama was quicker to embrace a range of Western innovations; for example, it had the first gas lighting system in Japan, the first railway and telegraph service (to Tōkyō), the first English-language newspaper (in 1861), the first red brick building, and the earliest local food industries catering to foreigners (dairies, breweries, bakeries, and so on).[32] Yokohama also moved more tea and silk to the United States than Kōbe, which had a larger Asian trade. And Yokohama had a more articulated trade network; for example, it opened shipping services to Bombay in 1893 and was the home of the Japanese-controlled NYK Lines (Nippon Yūsen Kaisha).[33] Yokohama was, in short, a generally more dynamic centre.

What can be said of Kōbe after its status as a treaty port ended? The shift from foreign to national and local administration encouraged the Meiji government to expand Kōbe's port infrastructure. A period of late-Meiji port expansion commenced in 1907 when plans were put in place to build the first Japanese floating dock, known as Shinko Pier. This port expansion activity lasted until 1922. In tandem with these infrastructure projects, Ōsaka's financiers were helping spur Kōbe's industrial growth by the end of the nineteenth century. Kōbe would experience higher growth than Yokohama thereafter and would top Yokohama in total import and export trade by 1914.

The forces of development unleashed on Kōbe's port in the early twentieth century were powerful. For instance, land reclamation along the port's shoreline before 1922 greatly expanded Kōbe's harbour capacity. By 1921, reclamation work in the bay had increased the harbour's freight-handling capacity by 2.1 million tons. One consequence of this was that the Bund – Kaigan-dori, the street that so self-consciously embodied the Concession's sense of place and grandeur – effectively ceased to be the "water street." When the shoreline was shifted seaward by some 100 metres, the Bund became an inland street; no longer did it offer the panorama that had so engaged early visitors. Today, Kaigan-dori is National Route No. 2, a wide motorway. The creation of new land by filling in the harbour continued with the massive expansion of Kōbe's portlands in 1973. This program created two large artificial islands, Port Island and Rokko Island, where multifunctional land uses have made for a remarkable urban landscape, one that combines container port installations with ultramodern housing, shopping, and institutional buildings. In the twenty-first century, trade through the port of Kōbe remains vital to Japan's export-led economy. At the time of the

Great Hanshin Earthquake in 1995, Kōbe was the most important container port in Japan. The earthquake and a protracted economic malaise in Japan have reshaped the internal ordering of port rankings within Japan, and on the world scale, Japan's ports have been outdistanced by other Asian ports in the past decade. Nevertheless, Kōbe remains an important port for foreign trade and has retained a vibrant and self-consciously cosmopolitan foreign community, one that exhibits a sense of history and belonging. Kōbe's port, then, maintains a legacy that rests on the heady days of 1868, when Westerners eagerly came ashore to inspect the Foreign Concession. In so doing they helped open a window on the West through the mechanism of the Foreign Concession that was Kōbe.

Glossary

Bakufu   The name given to the Tokugawa Shōgunate, also known as the Tokugawa Bakufu. The term literally means "tent office," or home of the general. The Shōgun's officials were collectively referred to as the Bakufu.

banto   A chief clerk, buyer, or intermediary in a Japanese-based trading firm or *hong*.

boo   A denomination of Japanese currency roughly equivalent to 3 boos = $1.00 in 1868.

cattie or catty   A unit of weight used in China and Japan, often in reference to the silk trade. It is equal to about $1^1/_3$ pounds, or about 0.67 kilogram.

comprador   A senior Chinese manager in a foreign company or bank.

daimyō   The powerful regional lords under the Tokugawa Shogunate. These men ruled the provinces from their large hereditary landholdings.

han   A fiefdom, estate, or domain under the Japanese feudal system. These were given to warriors of the daimyō.

hong   A Western-owned trading firm. Most of these firms started on the China coast, and several opened branches in Japan when it opened to trade. See also chapter 2, note 9.

karuma   In this reference, Isabella Bird is probably referring to a Japanese form of cart. It may have been a jinrikisha, or it may have been a cruder form of two-wheeled vehicle used to haul goods.

picul   An Asian unit of weight, often used in the silk trade, equal to 100 catties, or $133^1/_3$ pounds (about 60 kilograms)

Rokkō   The local name for the mountain range behind the City of Kōbe.

rōnin   Literally, "masterless samurai," that is, members of the warrior class who were cast adrift by the death or fall from grace of their master.

sendō A harbour boatman. It was these men who ferried Westerners to and from ships.

tsubo A common Japanese measure equal to 3.303 square metres, or 35.58 square feet. It was commonly used with regard to the area dimensions of house.

Explanatory Notes

Japanese name order   Customarily, the names of Japanese individuals are rendered with the family name preceding the given name. However, since this study focuses primarily on the foreign population, the Western naming pattern of family name last has been adopted.

A note on photographs   Most of the photographs found in the book have been cropped. Exposure and contrast have been digitally enhanced, watermarks and stains removed, and the final images rendered in black and white to emphasize the features being illustrated. The author has not edited or manipulated the images in any other way.

Notes

Preface

1 Milburn, *Oriental Commerce*.
2 Wallerstein, *The Modern World System*, vols. I–III.
3 Ogborn, *Indian Ink; Global Lives*.
4 Dejima was a small artificial island formed by cutting a channel through a peninsula in the harbour of Nagasaki. Originally built to house the Portuguese traders, the location continued to serve as the base for Dutch and Chinese trade from 1641 to 1853.
5 Kōbe Foreign Board of Trade, Proceedings, January to June 1868, unsourced clipping, Kōbe City Library, Historical Section, no date.
6 *Madame Butterfly*, first performed in 1904, is based on the short story of the same name written by John Luther Long and also on the experiences and writings of Pierre Loti, a French naval officer who was stationed at Nagasaki and there took a temporary bride. For an excellent description of the "Butterfly game" and the background to Loti's adventure in Nagasaki, see Barr, *The Deer Cry Pavilion*, 185–92.

1. Setting the Stage

1 Chaudhuri, *The Trading World; Trade and Civilization*; Prakash, *Dutch East India Company*; Chauduri, *Companies and Trade*.
2 See "Armenian Merchants of the Seventeenth and Early Eighteenth Centuries."
3 Cooper, "The Brits in Japan." For detailed studies of the English East India Company in Japan during this period, see Farrington, *The English Factory in Japan, 1613–1623*; and Massarella, *A World Elsewhere*.

4 Boxer, *The Christian Century in Japan, 1549–1650*.
5 Van Dyke, *The Canton Trade*.
6 Eng, "The Transformation of a Semi-Colonial Port City," 130–2.
7 Keith, *Constitutional History of India*.
8 Fairbank, *Trade and Diplomacy on the China Coast*, ch. 8.
9 Bowen, Lincoln, and Rigby, eds., *The Worlds of the East India Company*.
10 Sansom, *The Western World and Japan*, 169.
11 Yonemoto, "Maps and Metaphors."
12 The term Bakufu refers to the civil governing administrative circle of the Shōgunate.
13 Beasley, *The Modern History of Japan*.
14 Takekoshi, *The Economic Aspects*, ch. 75, 208–22.
15 For perhaps the best treatment of this crucial period, see Totman, *The Collapse of the Tokugawa Bakufu*; and *Early Modern Japan*, ch. 21.
16 Neither port was proximate to Yedo or Ōsaka, the principal population and economic centres of Tokugawa Japan. Shimoda was at the tip of the Izu Peninsula on the Pacific coast; Hakodate was at the southern tip of the northern island of Hokkaidō. Undoubtedly these ports were acceptable first steps for the Americans because they were among the closest landfalls for ships crossing the Pacific from North America.
17 Totman, *Early Modern Japan*, ch. 21; see also Sansom, *The Western World*, chs. 11 and 12.
18 Up until the Meiji Restoration, Tokyo was known as Edo, which was anglicized as Yedo. Given this book's emphasis on seeing this period through the lens of Western observers and actors, Yedo will be used hereafter.
19 The Meiji Restoration refers to a reconfiguration of the Japanese government as a result of which the emperor was restored to prominence as Japan's unifying and central authority after a long period of political marginalization under the Tokugawa Shōguns. The era began when the young Emperor Meiji assumed the throne in 1867. It saw the establishment of a highly centralized government. It was also marked by the establishment of equality of the Japanese people before the state and by the modernization of the economy and the education, banking, and legal systems.
20 Murphy, "On the Evolution of the Port City," in Broeze, *Brides of the Sea*, 242; Wheatley, *From Court to Capital*.
21 Ōsaka was included because it was Japan's second largest city as well as its most important economically. It was also the centre of Japan's "industrial" activity and had a large merchant class, who were growing in wealth and influence even though the Tokugawa regarded them as low in status. Indeed, it has been argued that the Bakufu had sought to keep Westerners

isolated from Ōsaka, in order to prevent the forces there from becoming even wealthier and more powerful. The delays in opening of Kōbe and Ōsaka, and the attempt to isolate Western interests in marginal ports such as Hakodate, Niigata, and Shimoneseki, were seen as part of this strategy. See Howe, *The Origin of the Japanese Trade Supremacy*, 76.

22 The rallying cry of the dissidents was *sono joie* (rough translation: "protect the Emperor; expel the barbarians"). The movement expressed itself in a small but widely heralded number of incidents in which "two-sworded" men – that is, samurai as well as masterless samurai or *ronin* – assaulted and murdered Westerners near Tokyo and Yokohama and also at Kōbe.

23 Fairbank, *Trade and Diplomacy on the China Coast*; Murphy, *Shanghai: Key to Modern China*; "Traditionalism and Colonialism."

24 Reeves, Broeze, and McPherson, "Studying the Asian Port City," in *Brides of the Sea*, 45.

25 Examples of Western studies of Japanese urbanism and its cities include: Gary Allinson, "Japanese Cities in the Industrial Era," *Journal of Urban History* 4 (1978): 443–75; David H. Kornhauser, *Urban Japan: Its Foundation and Growth* (London and New York: Longmans, 1976); Edward Seidensticker, *Low City, High City – Tokyo from Edo to the Earthquake, 1867–1923* (New York: Alfred A. Knopf, 1983); André Sorensen, *The Making of Urban Japan: Cities and Planning from Edo to the Twenty-First Century* (London: Nissan Institute / Routledge Japan Studies Series, 2002); James L. McClain and Osamu Wakita, eds., *Ōsaka: The Merchant's Capital of Early Modern Japan* (Ithaca: Cornell University Press, 1999); Paul Wheatley and Thomas See, *From Court to Capital: A Tentative Interpretation of the Origins of the Japanese Urban Tradition* (Chicago: University of Chicago Press, 1978); and Paul Waley and Nichola Fiévé, eds, *Japanese Capitals in Historical Perspective: Place, Power, and Memory in Kyōto, Edo, and Tokyo* (London: Routledge Curzon, 2003). There is a striking absence of studies by Japanese scholars in translation.

26 Hoare, *Japan's Treaty Ports and Foreign Settlements*.

27 Murphy, *The American Merchant Experience in Nineteenth Century Japan*.

28 Honjo, *Japan's Early Experience of Contract Management in the Treaty Ports*.

29 See Williams, *Tales of the Foreign Settlements in Japan*; *Shades of the Past*; See also a more personal reflection of being a foreigner in Japan, in Williams and Williams, *West Meets East*.

30 Toby, *State and Diplomacy in Early Modern Japan*; Sugiyama, *Japan's Industrialization in the World Economy*; Kazui and Videen, "Foreign Relations during the Edo Period."

31 Lewis, *Frontier Contact Between Choson Korea and Tokugawa Japan*; Innes, "The Door Ajar," cited in Lewis; Sakai, "The Satsuma-Ryukyu Trade."
32 See, for example, Weigend, "Some Elements in the Study of Port Geography."
33 Hornsby, "Discovering the Mercantile City in South Asia," 135.
34 McGee, *The Southeast Asian City*.
35 Kosambi and Brush, "Three Colonial Port Cities in India."
36 Vance, *The Continuing City*.
37 Bowden, "Growth of the Central District of Large Cities,"; "The Internal Structure of the Colonial Replica City"; "The Mercantile City in North America"; unpublished field guides.

2. The Creation of Kōbe's Foreign Concession

1 Estimates for the population of Hiōgo vary. Rutherford Alcock, the first British official to inspect the port, in 1861, thought that the population was perhaps 20,000 inhabitants. See Cortazzi, *Victorians in Japan*, 156.
2 Japan Chronicle, *History of Kōbe*, 4.
3 Japan Chronicle, *History of Kōbe*, 4.
4 Williams, *Tales of the Foreign Settlement in Japan*, 65.
5 Williams, *Tales of the Foreign Settlement in Japan*, 65.
6 British Foreign Office (hereafter FO), Correspondence from the British Consul at Hiōgo, Japan, Public Record Office, Kew, FO 262 148, 15 February 1868.
7 FO, 15 February 1868.
8 In the course of preparing this study it has been necessary to wrestle with a variety of place designations that are cause for some confusion. First, the existing urban settlement of Hiōgo constituted a sizeable city, which was centred to the west of the Minatogawa (Minato River). The City of Hiōgo, later spelled Hyōgo, was part of the new prefecture of Hiōgo created by the Meiji government in 1876 as an amalgamation of the former provinces of Harima, Tajima, and Awaji as well as parts of Tamba and Settsu. The area that was known as Kōbe at the time of foreign contact was an outlying village of the city of Hiōgo. When the foreign settlement was created on ground immediately abutting Kōbe, it became the practice among foreigners to refer to this existing settlement as the "Native Town" – a practice that mirrored the pattern of usage employed by the British and others in India and other colonial port cities in this era. As will be demonstrated later, the growth of the city of Hiōgo in the succeeding decades of the nineteenth century caused Kōbe and a number of the other outlying villages to

be annexed to the city, and in 1889 the city of Hiōgo was renamed Kōbe to end the confusion with respect to Hyōgo prefecture but also in recognition that the name Kōbe was increasingly being used in reference to the treaty port. It is perhaps a measure of the importance of the port, both within Japan and beyond, that the name Kōbe became the official name for the larger urban place that was taking shape by the last decade of the nineteenth century.

9 The term *hong* was employed in China and subsequently in Japan in reference to the major foreign trading firms that acted as importers/exporters, shipping agents, and processors of raw goods such as tea and cotton. Several of the larger firms had representation in several ports – for example, Jardine Matheson and Augustine Heard.

10 Augustine Heard Collection, Baker Library, Harvard University, Cambridge, MS 766 1835–1892 (hereafter AHC), case 27 f 56. The term *tsubo* refers to a common Japanese area measure. One *tsubo* equals 3.3 square metres or 35.58 square feet. The Rio, or *ryō*, was a Japanese currency value based on the weight of various minted metals: gold, silver, or copper. At the time of Perry's arrival in Japan, the exchange rate was one gold *ryō* to four Mexican silver dollars. Frost, *The Bakumatsu Currency Crisis*, 13.

11 Gambier, *Links in My Life on Land and Sea*, 372–3.

12 For more on Yokohama, see Hoare, *Japan's Treaty Ports and Foreign Settlements*, ch. 5.

13 Harold S. Williams Papers, National Library of Australia, Canberra, MS 6681/5. Quotes derived from the text of a lecture given to the Fellowship Luncheon Club entitled "So-Called Good Old Days."

14 Williams Papers, "So-Called Good Old Days." In the 1860s, Sakai was a small port at the mouth of the Yamamoto River west of the main urban concentration of Ōsaka. As noted, the effectiveness of Ōsaka's port was compromised by the Yamamoto's silting and sand bars, which rendered it difficult for Western-designed ships to enter the harbour. Today, Sakai houses the modern port of Ōsaka.

15 Williams Papers, "So-Called Good Old Days."

16 Williams Papers, "So-Called Good Old Days." This poem was in fact performed at an amateur dramatic event on 12 February 1868, six weeks after the port was officially opened.

17 Prince Hirobumi Itō (a.k.a. Shunske Itō) (1841–1909), the first governor of the newly created Hiōgo prefecture, was a distinguished figure in early Meiji Japan. Itō began life as a retainer of the lord of Chōshū and as a young man was sent on a secret mission to Yedo to report to his lord on the actions of the Shōgun's government. This visit caused Itō to turn his

attention seriously to the study of the British and other military systems. As a result he persuaded the *daimyō* of Chōshū to remodel his army. Later, in 1863, Itō with four other young men of the same rank risked their lives by committing the then capital offence of visiting a foreign country. From Nagasaki they obtained passages on a vessel that was about to sail for Shanghai. From there they travelled to London. When the western *daimyōs* challenged the Shōgun's accommodation of the Western powers, Itō hurried back to Japan. In 1868, Itō was made governor of Hiōgo – a move that he largely engineered himself during the chaotic days between the final collapse of the Shōgunate regime and the beginning of the Meiji Restoration. Over the following year he became Vice-Minister of Finance. In 1871 he and three other leading lights accompanied Plenipotentiary Ambassador Tomomi Iwakura on an important mission to Europe, which resulted in the adoption of Western military, naval, and educational systems. After his return to Japan, Itō served in several cabinets, and in 1886 he accepted the office of prime minister, a post he resigned in 1901. In September 1907 he was advanced to the rank of prince and became President of the Privy Council in Japan. In October 1909, while on a visit to Harbin, he was shot dead by a Korean assassin.

18 FO 262, 148, Correspondence from Mr. Crowder to Foreign Office, 1 February 1868.
19 FO 262, 148. Contingency plans were hastily made to evacuate all foreigners to ships in the harbour if necessary. Myburgh judged the situation at Hiōgo to be of low risk owing to the limited strategic advantage to be gained by the rebels. Members of the staff based at Ōsaka, on the other hand, were caught in a more dangerous situation and did have to evacuate.
20 Hart advertised his professional practice in the port on 26 August 1868 in the *Hiōgo and Ōsaka Herald*. It seems likely that he was in the port earlier that this date, directing preparatory work on the site of the settlement. It is also likely that he had been in Shanghai prior to arriving in Japan.
21 Stanislawski, "The Origin and Spread of the Grid-Pattern Town."
22 Reps, *The Making of Urban America*.
23 The wall surrounding the Chinese city was apparently constructed in the 1860s, thereby inserting a clear barrier between the Chinese and foreign communities. Rowe, *Hankow*, 71.
24 Known variously as the Public Garden, the Recreation Ground, and Huangpu Park, it was situated adjacent to the British Consulate on reclaimed land. From the beginning it invited controversy in that the Chinese were not permitted to use the park. See Vickers and Wasserstrom, "Shanghai's 'Dogs and Chinese Not Admitted' Sign."

25 The term Bund is one of a number of terms that became part of a lexicon of the Asian trading world. It was borrowed from Hindi/Urdu and referred to an embankment or waterfront. Within the Asian trading world it was applied to the streets or quays that fronted several treaty port settlements along the harbour shoreline.
26 FO, March 1868.
27 By convention, British road surveys of this time used the measure of 1 chain, or 66 feet, as a standard road allowance. It seems apparent that road allowances in this case were set at one half of this dimension.
28 The standard literature uses the term Foreign Powers to refer to the heads of the legations that had established treaties with Japan following the opening of Japan. These were Great Britain, France, Prussia, Russia, and the United States.
29 FO 262, 148, 23 March 1868. The memorandum was dated 20 March 1868. The signators, in person or by proxy, were L.B. Glover, F. Blake, Eug Van der Heyde, R.D. Robison, and Paul Heinemann, all of them representing trading firms established in Nagasaki or Yokohama, or both.
30 FO 262, 148, 30 March 1868.
31 FO 262, 148, 26 August 1868. It is not clear to whom Lowden was referring. As we have seen, Heard & Co. was one firm that had acquired such a favourable location in the Native Town in July 1868. Nevertheless, the firm also purchased lot 7 in the Concession when the auction was held on 10 September. The firm maintained the original property in the Native Town at least until 1874, when it sought to sell it, without apparent success. See AHC, case 27, f 56 Property – Japan – Kōbe 1868–74.
32 The dollar referred to here was the Mexican silver dollar, which served as the principal trade currency used by Westerners in Asia during this period.
33 FO 262, 148, 18 September 1868.
34 FO 262, 148, 18 September 1868.
35 Kōbe City Archives, #61:1, *Japan Gazette Hong List and Directory*.
36 Murata, *Kōbe Kaiko Sanjunenshi*; *Hiōgo News*, 18 May 1870. Three or four purchasers of lots apparently defaulted, and these lots were finally sold in 1873.
37 Williams, "Shade of the Past."
38 FO 46 158.
39 FO 46 158.
40 The term godown evidently came to be used in Malaya as well. In 1998 the author photographed a building in the old port town of Melaka (a.k.a. Malacca) that carried a sign indicating it was a "godown."
41 *Hiōgo News*, 4 February 1869.

42 *Hiōgo News*, 27 July 1869.
43 Reconstructed from a diagram in the AHC, LV–21, folio 21, Correspondence from Blake to Albert Heard, 26 April 1870.

3. Establishing Municipal Government and Services in the Concession

1 Hoare, *Japan's Treaty Ports and Foreign Settlements*.
2 British Foreign Office (hereafter FO), Correspondence from the British Consul at Hiōgo, Japan, Public Record Office, Kew, FO 262 148, 18 September 1868.
3 FO 262, 148, 13 October 1868. The dispatch detailing the basis of municipal governance was prepared by Russell Robertson for Sir Harry Parkes, the British minister. In it he uses the term "land renters" when clearly referring to those who had purchased lots. The term renter presumably refers to the fact that lot owners paid an annual "ground rent."
4 Japan Chronicle, *History of Kōbe*, 21.
5 FO 262, 148, Parkes to Lowder, 23 October 1868.
6 Japan Chronicle, *History of Kōbe*, 21.
7 *Hiōgo and Ōsaka Herald*, 20 January 1869.
8 *Hiōgo and Ōsaka Herald*, 16 January 1869
9 *Hiōgo and Ōsaka Herald*, 16 January 1869.
10 *Hiōgo and Ōsaka Herald*, 17 February 1869.
11 *Hiōgo and Ōsaka Herald*, 18 June 1869.
12 *Hiōgo and Ōsaka Herald*, 23 June 1869.
13 *Hiōgo and Ōsaka Herald*, 10 August 1869.
14 *Hiōgo and Ōsaka Herald*, 21 August 1869.
15 *Hiōgo and Ōsaka Herald*, 4 September 1869.
16 *Hiōgo and Ōsaka Herald*, 5 June 1869.
17 *Hiōgo and Ōsaka Herald*, 19 June 1869.
18 *Hiōgo and Ōsaka Herald*, 24 July 1869.
19 Mechanic's Institutes were educational institutions set up to provide technical instruction for men. Begun in Edinburgh in the 1820s, they became popular throughout Britain, the United States, and Canada, particularly because they provided access to what were in effect public libraries.
20 *Hiōgo News*, 22 January 1870.
21 *Hiogo News*, 22 January 1870.
22 *Hiōgo News*, 22 January 1870
23 A *koku* of rice was equal to 150 kilograms and was historically the amount needed to feed an individual for a year.

24 *Hiōgo News*, 7 September 1888.
25 *Hiōgo News*, 22 January 1870.
26 *Hiōgo News*, 7 September 1870.
27 FO 262, 231, Adams to Gower, 28 December 1871.
28 FO 262, 236/R.103, C. Shepard to R.G. Watson, n.d. The correspondence contains a memo by Shepard summarizing his protest, dated 27 August 1872.
29 FO 262,246, Gower to Watson, no. 15, 18 February 1873.
30 Hoare, *Japan's Treaty Ports and Foreign Settlements*, 123.
31 *Hiōgo News*, 26 and 29 January 1876.
32 *Hiōgo News*, 15 March 1876.
33 Hoare, *Japan's Treaty Ports and Foreign Settlements*, 123.
34 *Hiōgo News*, 19 January 1870.
35 *Hiōgo and Ōsaka Herald*, 9 January 1869.
36 *Hiōgo News*, 2 July 1870.
37 A *rōnin* was a member of the warrior class who had been cast adrift by the death or fall from grace of his master. A number of such individuals had joined marauding bands, which threatened Westerners during the confused period between the fall of the Tokugawa regime and the installation of the Emperor Meiji.
38 FO 262, 149/52.
39 FO 262, 149 /52.
40 FO 262, 150/ 40.
41 *Hiōgo and Ōsaka Herald*, 3 April 1869.
42 *Hiōgo and Ōsaka Herald*, 28 April 1869.
43 *Hiōgo and Ōsaka Herald*, 20 February 1869.
44 *Hiōgo and Ōsaka Herald*, 20 April 1869.
45 *Hiōgo and Ōsaka Herald*, 23 June 1869.
46 *Hiōgo and Ōsaka Herald*, 26 January 1869.
47 *Hiōgo News*, 12 February 1870, refers to a break-in at Messrs. Wilkins and Robison's godown in the Native Town.
48 *Hiōgo and Ōsaka News*, 16 February 1869.
49 *Hiōgo News*, 8 June 1870.
50 Yumiko Yamamori, *Japanese Export Furniture with Particular Emphasis on the Meiji Era (1868–1912)*, Web publication accompanying a digital art exhibition of Japanese arts, http://www.euronet.nl/users/artnv/main.html.
51 *Hiōgo and Ōsaka Herald*, 13 January 1869.
52 *Hiōgo and Ōsaka Herald*, 16 January 1869.
53 *Hiōgo News*, 26 January 1870.
54 *Hiōgo and Ōsaka Herald*, 6 January 1881.
55 *Hiōgo and Ōsaka Herald*, 20 January 1869.

56 *Hiōgo and Ōsaka Herald*, 21 March 1871.
57 FO 262, 216, March 1872.
58 FO 262, 216, May 1871.
59 FO 262, 216, 249–98.
60 FO 262, 216, May 1871, 337, 421.
61 *Hiōgo News*, 3 February 1881.
62 For a description of the brothels at Kōbe, see Abend, *Treaty Ports*, 178–9.
63 *Hiōgo News*, 17 April 1879.
64 *Hiōgo News*, 15 January 1886.
65 *Kōbe Chronicle*, 25 January 1898.
66 FO 262 148, Lowden to Parks, 5 September 1868.
67 *Hiōgo and Ōsaka Herald*, 11 September 1869.
68 *Hiōgo and Ōsaka Herald*, 4 September 1869.
69 *Hiōgo and Ōsaka Herald*, 11 September 1869.
70 *Hiōgo News*, 29 December 1869.
71 *Hiōgo News*, 12 February 1870.
72 *Hiōgo News*, 4 January 1881.
73 *Hiōgo News*, 15 January 1886.
74 *Kōbe Chronicle*, 18 January 1898.
75 *Kōbe Chronicle*, 12 March 1898.
76 *Kōbe Chronicle*, 30 March 1898. See also FO, 262 148, letter, Governor Itō to Parks, 30 March 1868.
77 Hoare, *Japan's Treaty Ports and Foreign Settlements*, 120.
78 FO 262 148, Myburgh to Parks, 13 January 1869.
79 *Hiōgo and Ōsaka Herald*, 12 June 1869.
80 *Hiōgo and Ōsaka Herald*, 12 June 1869; Williams suggests, incorrectly, I believe, that the date for the founding of this hospital was 1871. However, he was undoubtedly correct in noting that it soon became apparent that the house at Ikuta-mae would not be suitable in the years ahead as that part of the city was becoming more and more congested with small shops; above all, the hospital's supply of good water had become seriously endangered. In 1874 a more suitable lot was leased from Japanese owners at Yamamoto-dori 1-chome, near Ichinomiya Shrine. There, with money raised from the members, a new hospital was built, although in truth it was only a two-storey building. Nevertheless, it served its purpose for more than thirty years, until it no longer proved adequate. Difficulties were eventually encountered when the Japanese owner of the land sought to evict the hospital and recover possession of the land, in spite of the terms of the lease. The location was saved only after a long and hard fight

by the British Consul. See Harold S. Williams Papers, National Library of Australia, Canberra, MS 6681/7, International Hospital.
81 Hoare, "The Japanese Treaty Ports 1868–1899," 257.

4. Forging an Economy

1 *British Parliamentary Papers "Report 1868,"* IV:23.
2 Sugiyama, *Japan's Industrialization in the World Economy, 1859–1899*, 45.
3 Takekoshi, *The Economic Aspects of the History of the Civilization of Japan*, III:219. This process had been occurring during much of the later Tokugawa period but was particularly pronounced during the Tempo period, 1830–43. See also Frost, *The Bakumatsu Currency Crisis*, 11.
4 Yoshihara, *Japanese Economic Development*, 2.
5 The term "Mexican silver dollar" came to be used as a generic term for the "trade dollar" coins produced by several countries. There were, in fact, a variety of trade dollar coins in circulation, including the British crown, the German Reich five mark coin, the Peruvian "un sol," the Italian five lire, the Dutch 2½ gulden, the late-nineteenth-century Spanish five peseta, and, earlier, the eight reales with the head of the Spanish king. In the late nineteenth century, specific trade dollars were produced. The British had a "one dollar" coin with Britannia and, on the reverse, the denomination in Chinese and Jawi Malay. The United States produced coins with the Liberty head or Eagle that bore the text "one troy ounce silver trade unit, 0.999 pure silver." A "One Tael" coin with the British coat of arms was struck for trade in Shanghai and Hong Kong. The French produced an "Indo-Chine Française Piastre de Commerce." Allen, *The Encyclopedia of Money*.
6 A British experiment in manufacturing silver dollars at Hong Kong had been abandoned, and the Hong Kong Mint's machinery was for sale. This was bought by Japan, and the equipment and British staff relocated to Ōsaka, where the Imperial Mint was to make its home. This was nominally a Japanese institution; however, it was run by foreigners, and this fact was further cause for tension during the early years of its existence.
7 Yoshihara, *Japanese Economic Development*, 3.
8 *Hiōgo and Ōsaka Herald*, 6 January 1869.
9 In 1863, Aspinall-Cornes secured the agency of the P&O steamship line, and this prestigious appointment led to the acquisition of brokerage agencies for Universal Marine, London & Oriental, Commercial Union, and Queen insurance companies. In 1868, Aspinall-Cornes was appointed the

first Lloyd's agent in Japan – a position that the company retained through the twentieth century.
10 *Hiōgo and Ōsaka Herald,* 10 February 1869.
11 Patterns of ship movement have been reconstructed from a detailed summary transcription of the ship arrivals published in the *Hiōgo and Ōsaka Herald* at the end of December 1869.
12 The *Cadiz* was an iron screw steamer rated at 816 tons and built in 1853.
13 The *Formosa* was a 637-ton iron passenger liner built in 1852 and employed on the P&O's Calcutta–Far East route and later on coastal services from Hong Kong.
14 The patterns of ownership of the ships employed in the sea lanes of Japan and China at this time reveal a remarkable fluidity. For example, the *Bahama* was built by Pierce and Lockwood at Stockton-on-Tees in 1862 and served first as a supply ship for Confederate raiders during the American Civil War. It arrived in Japan consigned to Dent & Co. in May 1864 and within a few months had been bought by the firm of Glover & Co. of Nagasaki, who within days sold it to the Kishu-han, who renamed it *Meiko Maru*. In 1868 she was acquired by Adrian & Co., who reverted her to her old name and British flag. After extensive repairs in Whampo (Shanghai), she was set to work the China–Japan trade. In April 1870 she arrived in Hiōgo, where she lay until July, when she was sold to the Kishu-han again and given back her Japanese name. By 1875, after a number of other moves, she had been acquired by Mitsubishi Jubin. The research on this and on several other vessels is attributed to T.W. Milne (1964), and was uncovered in the Harold Williams Papers, NLA MS 6681, Series 1(b)/12.
15 Barques were the workhorses of the so-called Golden Age of Sail in the mid-nineteenth century. Typically built of wood and with three or more masts, they carried fore-and-aft sails on the aftermost mast and square sails on all other masts. Because there were fewer full-rigged masts, barques were able to nearly match the performance of full-rigged ships but with smaller crews.
16 The opening of the Suez Canal in 1869 took more than 3,500 nautical miles (6475 kilometres) off the distance from London to Kōbe compared to the previous route via the Cape of Good Hope. The number of merchant steamers passing through the canal increased rapidly, from 1,042 in 1876 to 1,534 in 1880 and 2,514 in 1885. This significantly altered the speed of service offered by shipping companies such as the P&O line, the French Messageries Impériales, and those that followed, such as the Blue Funnel, Castle, and Glen lines, which entered the East Asia trade in the 1860s and 1870s.

17 Hearn, *Tracks in the Sea*.
18 Wray, *Mitsubishi and the NYK*, 4.
19 Wray, *Mitsubishi and the NYK*, 4.
20 Sasaki, "The Introduction of European-styled Vessels in Japan," 50.
21 Sasaki, "The Introduction of European-styled Vessels in Japan," 52.
22 Sasaki, "The Introduction of European-styled Vessels in Japan," 47.
23 Wray, *Mitsubishi and the NYK*, 4.
24 Wray, *Mitsubishi and the NYK*, 4.
25 British Foreign Office (hereafter FO), Letters to Consular Office in Kōbe and Yokohama, Public Record Office, Kew, 262 231, 15 February 1868.
26 *Hiōgo and Ōsaka Herald*, 20 February 1869; Report of the Meeting of the Hiōgo and Ōsaka Chamber of Commerce, 15 February 1869.
27 *Hiōgo and Ōsaka Herald*, 15 February 1869.
28 *Hiōgo and Ōsaka Herald*, Chamber of Commerce of Hiōgo and Ōsaka, Report from A.O. Gay, H.R. Mackenzie, A.J. Bauduin, J.S. Blydenburgh, H. St J. Browne, 27 July 1869.
29 National Archives and Records Administration (hereafter NARA), Washington, D.C., Microfilm 460, reel 1, 0177. Chamber of Commerce of Hiōgo and Ōsaka, Report of the Special Committee of the Hiōgo and Ōsaka Chamber of Commerce on the proposed Revisions to the Tariff on Tea and Silk, October 1868.
30 The report names a number of silk districts that cannot be located due either to poor rendering of the place name or because the geographic reference has disappeared. For example, Aechesan, Maggerhama, Mashtah, Sodai, Tamba, and Shida were said to be the principal silk-producing districts in the region accessible to Kōbe–Ōsaka. It is probable that these places were villages that generally lay in the upland valleys running from Nara in the southeast to Ōtsu in the northeast.
31 NARA, Microfilm 460, reel 1, 0177. Chamber of Commerce of Hiōgo and Ōsaka, Report of the Special Committee on the proposed Revisions to the Tariff on Tea and Silk, October 1868.
32 An excellent treatment of this topic is in Frost, *The Bakumatsu Currency Crisis*.
33 Report of the Chamber of Commerce of Hiōgo and Ōsaka, 1868.
34 The *hongs* in Kōbe, in addition to Western employees, had Chinese employees, who were valued by virtue of their familiarity with the procedures of commodity trade and in particular with the handling of money, and who could also act as go-betweens with the Japanese because they were far better at mastering the Japanese language than their Western bosses. The hierarchy of these Chinese employees extended from

compradors – trade brokers – to other positions such as *bantos* and *shroffs* (see following note), who counted the money and bartered exchange rates. Kuang Yung Pao, "The Comprador."
35 *Hiōgo and Ōsaka Herald*, 6 March 1869. The term *shroff*, derived from the Hindi *śarāf* and the Urdu *sharāf*, became widely used in the Far East trade.
36 NARA, Microfilm 460, Reel 1, U.S. Consular Record of Hiōgo and Ōsaka, Commercial Report, Daniel Turner to Cadwallader, 15 February 1875.
37 NARA, Microfilm 460, Reel 1, 0812, U.S. Consular Record of Hiōgo and Ōsaka, 3 August 1876.
38 Chamber of Commerce of Hiōgo and Ōsaka, Report, October 1868.

5. Finding a Mercantile Staple for Kōbe

1 Federico, *An Economic History of the Silk Industry*; Ma, "The Great Silk Exchange; idem, "The Modern Silk Road"; Li, "Silks by Sea."
2 Sugiyama, *Japan's Industrialization in the World Economy*, 79, 86, 94.
3 This activity, while centred on Spitalfield, also included large portions of Bethnal Green, Shoreditch, Whitechapel, and Mile End New Town. See Thirsk, *Alternative Agriculture*.
4 Matsui, *The History of the Silk Industry in the United States*; Margrave, *The Emigration of Silk Workers*.
5 Sugiyama, *Japan's Industrialization in the World Economy*.
6 Sugiyama, *Japan's Industrialization in the World Economy*, 111–12.
7 *Brief History of the Japanese Raw Silk Industry*, http://www.fasid.or.jp/public/paper3/e-2.html, accessed 17 July 2002.
8 Federico, *An Economic History of the Silk Industry*, 36–42.
9 Wittner, *Technology and the Culture of Progress in Meiji Japan*, 66. The Italians developed machines to spin the silk yarns and these were sought out by producers in England and France as well. The silk industry in Italy had long been centred in the north, on Florence and Venice and also around Lake Como.
10 Sugiyama, *Japan's Industrialization in the World Economy*, 112.
11 Sugiyama, *Japan's Industrialization in the World Economy*, 112.
12 Sugiyama, *Japan's Industrialization in the World Economy*, 91.
13 *Brief History of the Japanese Raw Silk Industry*.
14 Sugiyama, *Japan's Industrialization in the World Economy*, 113–25.
15 *British Parliamentary Papers*, "Report 1890," VIII:4.
16 The telegraph arrived in Japan in 1872. This entailed a complex routing of lines: a trans-Siberian traverse from Russia, then under the Baltic to

Denmark, and thence to other European locations and finally to North America via the Atlantic undersea cable. While this technology was available to both Western and Japanese merchants, the latter may have been slower to take up the possibilities owing to the complexities of adapting telegraphy to Japanese written characters. See Itoh, "The Danish Monopoly on Telegraph In Japan."
17 Sugiyama, *Japan's Industrialization in the World Economy*.
18 Davies, *The Business, Life and Letters of Frederick Cornes*.
19 *British Parliamentary Papers*,Report," IV:22–5.
20 *British Parliamentary Papers*, "Report, 1868," VI:250.
21 *Hiōgo and Ōsaka Herald*, 20 February 1869, Report of the Meeting of the Hiogo and Osaka Chamber of Commerce, 15 February 1869.
22 British Foreign Office (hereafter FO), Correspondence from the British Consul at Hiōgo, Japan, January–June 1868, Public Record Office, Kew, FO 262 148, #40, 28 January 1871.
23 FO 262 148, #40, 28 January 1871.
24 *British Parliamentary Papers*, "Report 1870," "Report for the Year 1870 on the Trade of Hiōgo and Ōsaka": IV:28.
25 Statistics compiled by the Hiōgo and Ōsaka Chamber of Commerce for 1879.
26 *British Parliamentary Papers*, "Report,1879" VI:7.
27 *British Parliamentary Papers*, "Report, 1884," VII:18.
28 *British Parliamentary Papers*, "Report, 1886," VII:3.
29 *Hiōgo News*, 6 November 1888.
30 *British Parliamentary Papers*, "Report 1893," 3; IX:3
31 Sugiyama, *Japan's Industrialization in the World Economy*, 140–1.
32 Sugiyama, *Japan's Industrialization in the World Economy*, 140–1.
33 Sugiyama, *Japan's Industrialization in the World Economy*, 140–1.
34 Augustine Heard Collection, Baker Library, Harvard University, Boston, (hereafter AHC), case 27, f 56, Property – Japan – Kōbe 1868–74. Agreement, 6 July 1868, between George Fairly Heard and Kuromeyo Yoshi.
35 AHC, Franklin Blake to G.F. Heard, 10 February 1871.
36 AHC, HM-23-3, H.G. Bridges to A.F. Heard, 28 April 1873.
37 AHC, Bridges to A.F. Heard, 28 April 1873.
38 AHC, Bridges to A.F. Heard, 28 April 1873.
39 AHC, Bridges to A.F. Heard, 28 April 1873 and 4 July 1873.
40 AHC, Bridges to A.F. Heard, 22 November 1873.
41 National Archives and Records Administration (hereafter NARA), Washington, DC, Microfilm 460, Reel 3, 1886, August 1886.

42 For images of tea producers and the harvest and preparation processes, see http://www.baxleystamps.com/litho/ta/macy_tea_color.shtml, viewed 24 August 2012. This website reproduces a photo album of forty hand-coloured collotype images, ca. 1890–1910, and carries the inscription "With Compliments of Geo. H. Macy & Co." Macy had an important tea export operation in Japan during this period. It is noteworthy that the T. Eaton Co., the leading Canadian department store chain, had its own agents in Japan at this time.
43 Terry, *Guide to the Japanese Empire*.
44 NARA, Microfilm 460 Reel 2, U.S. Consular Reports, 1877, 10 October 1877.
45 NARA, Microfilm 460, Reel 3, U.S. Consular Reports, 1883, Stahel to Adee, 26 May 1883.
46 NARA, Microfilm 460, Reel 2, U.S. Consular Report, 1880, Stahel to Payson, 28 September 1880.
47 *Hiōgo News*, 25 September 1888.
48 In 1880 it was reported that four-fifths of tea shipments have been conveyed in British or German vessels, with American ships carrying the other one-fifth. NARA, Reel 2, 29 January 1880.
49 *British Parliamentary Papers*, "Report 1879," VI:7. See also Webb, "Shipping Light."
50 NARA, Microfilm 460, Reel 2, U.S. Consular Report, 1879, Stahel (writing from Washington while on two months' leave) to Payson, 15 February 1881.
51 NARA, Microfilm 460, Reel 2, U.S. Consular Report, 1881, Stahel to Payson, 10 June 1881.
52 Sugiyama, *Japan's Industrialization in the World Economy*, 151.
53 *British Parliamentary Papers*, "Report, 1889," VII:4.
54 *British Parliamentary Papers*, "Report, 1892," IX:18.
55 *British Parliamentary Papers*, "Report, 1892," IX:18.
56 Sugiyama, *Japan's Industrialization in the World Economy*, 151.
57 *British Parliamentary Papers*, "Report, 1881," VII:13.
58 *British Parliamentary Papers*, "Report, 1883," VII:6.
59 *The Canadian Encyclopedia*, vol. 1 (Edmonton: Hurtig, 1985), 277.
60 *British Parliamentary Papers*, "Report 1887," VIII:3.

6. The Morphology of the Settlement and the Development of a Pleasing Townscape

1 *Hiōgo News*, 4 March 1870.
2 *Hiōgo News*, 6 July 1870.

3 National Archives and Records Administration (hereafter NARA), Microfilm 460, reel 3, U.S. Consular Report, 1884, 0452.
4 *Hiōgo News*, 20 January 1881.
5 *Hiōgo News*, 20 January 1881.
6 The Kōbe Regatta and Athletic Club was created in 1870 at a time when the Euro-American population of the port numbered but 271.
7 NARA, Microfilm 460, Reel 3, U.S. Consular Report, 1884, 0357, 24 November 1884.
8 *British Parliamentary Papers*, "Report 1887," VIII:6.
9 *British Parliamentary Papers*, "Report 1887," VIII:6.
10 *Hiōgo News*, 6 July 1888.
11 Cortazzi, *Victorians in Japan*, 162–3.
12 The Japanese term *jinrikisha*, from which the more commonly heard term "rickshaw" derives, was a form of two-wheeled conveyance for transporting people. It was pulled by a runner or driver. These vehicles became common in many parts of Asia in the nineteenth century, and the term *jinririksha* was used in Singapore in reference to the Jinrikisha Station. Cortazzi, *Victorians in Japan*, 163.
13 Cortazzi, *Victorians in Japan*, 162.
14 For a deeper exploration of this process, see Jones, *Live Machines*; and Burks, *The Modernizers*.
15 *Hiōgo News*, 28 July 1871.
16 Harold S. Williams Papers, National Library of Australia, Canberra, MS 6681, Series 1(b), folder 37. See also Japan Chronicle, *History of Kōbe*, 22. The latter account suggests that the decision to locate the route farther down the Hill was determined by the prospect that the popular Horse Racing Course would have to be relocated, a step that ultimately proved too difficult to overcome.
17 NARA, Microfilm 460, Reel 1, U.S Consular Report, 1876, 0044, Newitter to Chandler, 31 July 1876.
18 NARA, Microfilm 460, Reel 1, U.S Consular Report, 1876, 0495.
19 NARA, Microfilm 460, Reel 1, U.S Consular Report, 1876, 0049.
20 NARA, Microfilm 460, Reel 1, U.S Consular Report, 1876, 0895, Newitter to Cadwallader, 3 March 1877.
21 *Hiōgo News*, 5 October 1888.
22 *Hiōgo News*, 6 November 1888.
23 *A Centennial Tribute to Kōbe*, 11.
24 Williams Papers, series 16, folder 34. Today this street is called Flower Road and forms one of the main arterial streets of the central business district.

208   Notes to pages 120–32

25 This analysis is based on reconstructing owner/occupancy data from a variety of sources including newspaper advertisements, electoral rolls, and directories. There are sizeable gaps in that no data emerges for some lots, many of which must have been employed for the construction of godowns.
26 Rudyard Kipling, 1889, as cited in Cortazzi. *Victorians in Japan*, 166.
27 *Hiōgo News*, 4 January 1881.
28 Williams Papers, Box 16 folio A-C: 2.
29 Japan Chronicle, *History of Kōbe*, 29.
30 *Hiōgo News*, 4 February 1869.
31 *Hiōgo News*, 18 June 1868.
32 Japan Chronicle, *History of Kōbe*, 30.
33 *Hiōgo News*, 3 January 1884.
34 *Hiōgo News*, 7 September 1888.
35 *Hiōgo News*, 20 October 1888.
36 *Hiōgo News*, 12 November 1888. It was later reported by the *Hiōgo News* on 17 November that "some rascal has been cutting the wires" – an indication that the advance of modernity or the failure of Kobe Electric to accede to the council's wishes was precipitating guerrilla action.
37 Moto-machi was an important east–west street near the rear of the Native Town; it formed the continuation of Ura-machi, which was the northerly bounding street of the Concession.
38 *Hiōgo News*, 20 November 1888.
39 Heubner, "Architecture and History in Shanghai's Central District."
40 Japan Chronicle, *History of Kōbe*, 22.
41 Heubner, "Architecture and History in Shanghai's Central District," 212.
42 Author's observations.
43 *Hiōgo News*, 1 September 1888.
44 The periodic electoral lists published in the port's newspapers invariably provided the residential lot number of the individuals named. For example, from the list for 1876 we learn that C.B. Bernard had the same address as H.St.J. Browne, L.R. Goldsmith, and M.T.R. Macpherson, all British nationals apparently residing in business premises of Browne & Co. *Hiōgo News*, 20 December 1876.
45 A large literature exists on American domestic housing. Typical of this genre are Virginia McAlester and Lee McAlester, *A Field Guide to American Houses* (New York: Holt, 1884); and William Morgan, *The Abrams Guide to American House Styles* (New York: Abrams Books, 2008).
46 Finn, *Meiji Revisited*, 66.
47 Yip, "Neo-Colonial Treaty Port Cities on the China Coast."

48 Prior to coming to Hiōgo, Hart had been in Shanghai from at least 1865. As noted in chapter 2, Hart prepared the first Westerners' plan and map for the Concession and was tasked with laying out the townsite. He subsequently became the Municipal Council's first civil engineer. Williams Papers, Series 1(b)/file 35.
49 *Hiōgo News*, 27 November 1888.
50 *Hiōgo News*, 27 November 1888.
51 Finn, *Meiji Revisited*, 68.
52 "Bungalow" has come to refer to the modest one- or storey-and-a-half dwelling in North America and elsewhere. Originally, however, the term referred to a housing style originating with the Bengali *bangaloo*. This hipped-roof dwelling with exterior verandas was modified to suit Western sensibilities by the British in India. It was particularly evident in the "hill stations" and retreats developed by the British and other Europeans in South Asia, and it eventually spread as a housing type from there to Britain and other settler societies, such as Australia and indeed North America. See King, *The Bungalow*; and Aiken, *Imperial Belvederes*.
53 Williams Papers, MS 6681, Series 16, folder 34, 3.
54 *Hiōgo News*, 9 April 1871.
55 Bird, *Unbeaten Tracks in Japan*, II:216.
56 Alcock, *The Capital of the Tycoon*, II:102.
57 *A Centennial Tribute of Kōbe*. 21.
58 *Hiōgo News*, 18 June 1868.
59 It appears that the Hiōgo Hotel later relocated to the Native Town's Bund, as seen in Figure 6.8. This highlights the fluidity and footloose character of some businesses, particularly in the first days of the port's opening.
60 *Hiōgo News*, 4 February 1869.
61 *Hiōgo News*, 18 June 1868.

7. Life at the End of the World

1 Nation Archives and Records Administration (hereafter NARA), Microfilm 460, Reel 2, U.S. Consular Record, 0594, 11 October 1880.
2 NARA, Microfilm 460, Reel 2, U.S. Consular Report, 1880, 0594, 11 October 1880.
3 NARA, Microfilm 460, Reel 2, U.S. Consular Report, 1880, 0594, 11 October 1880.
4 NARA, Microfilm 460, Reel 2, U.S. Consular Report, 1878, 8 May 1878.
5 NARA, Microfilm 460, Reel 2, U.S. Consular Report, 1877, 0594, 25 September 1877.

6 NARA, Microfilm 460, Reel 3, U.S. Consular Report, 1885, 0659; 28 September 1885.
7 NARA, Microfilm 460, Reel 3, U.S. Consular Report, 1885, 0698;19 October 1885.
8 NARA, Microfilm 460, Reel 4, U.S. Consular Report, 1890, 0431, 8 September 1890.
9 NARA, Microfilm 460, Reel 2, U.S. Consular Report, 1881, 0431, 12 February 1881.
10 NARA, Microfilm 460, Reel 4, U.S. Consular Report, 1893, 0793, 26 January 1893.
11 NARA, Microfilm 460, Reel 4, U.S. Consular Report, 1888, 6 September 1888.
12 On 11 September 1877 the consul reported the death of Charles G. Henderson of Charleston, S.C., who died at International Hospital age twenty-nine. Henderson arrived in November 1871, and had found work as a clerk to a ship's chandler. Some months earlier he had suffered apoplexy, and he never recovered. He was found to be destitute, which required that funeral costs be covered by citizens. U.S. Consular Record, 11 September 1877.
13 From the age of twenty-two, James Favre-Brandt was an early (1863) and long-standing resident of Japan. His trading firm, which he operated in partnership with his brother Charles, had offices in both Yokohama and Ōsaka and specialized in the importing of watches, clocks, firearms, and military supplies. The firm acted as the consul for Switzerland in Ōsaka and Kōbe.
14 NARA, Microfilm 460, Reel 2, U.S. Consular Report, 1882, 0902, 8 August 1882.
15 NARA, Microfilm 460, Reel 3, U.S. Consular Report,1884, 18 January 1884.
16 For a discussion of these impulses as they related to American merchants residing in Japan, see Murphy, *The American Merchant*. See also Mangan and Walvin, *Manliness and Morality*; and Tosh, *Manliness and Masculinities*.
17 Honjo, *Japan's Early Experience of Contract Management*, 15; Barr, *The Deer Cry Pavilion*, 180. Various names were given to this district, including the Yoshiwara – a kind of generic usage that took its inspiration from Tokyo's famous pleasure district.
18 Hugill, *Sailortown*, 303.
19 Abend, *Treaty Ports*, 178.
20 Hugill, *Sailortown*, 307–8.
21 NARA, Microfilm 460, Reel 5, U.S. Consular Report, 1896, 0290, 29 April 1896.

22 Unfortunately the record also reports "the death of Prof. George C. Foulk, suddenly in the mountains near Miyanoshito, on Aug 6, where he had gone for failing health." Formerly a naval officer, he was U.S. Chargé d'Affairs in Korea in 1888 and was employed at a missionary college in Kyōto. He left a Japanese widow. NARA, Microfilm 460, Reel 4, U.S. Consular Report, 1894, 0859, 6 October 1894.
23 NARA, Microfilm 460, Reel 4, U.S. Consular Report 1887, 01020, 10 September 1887; and Reel 4, 6339, 14 October 1891.
24 NARA, Microfilm 460, Reel 5, U.S. Consular Report 1898, 14 April 1898.
25 Tamura, "Mary Kawatani Kirby, in Tamura, *Forever Foreign*, 29–44.
26 In the course of doing the research the author came into chance contact with individuals doing family genealogical research. In both instances the descendants revealed that they traced their roots to Scottish men living in Kōbe during this period. Both also revealed that these men formed relationships with Japanese women and produced children. In the case of one family, the Eurasian son eventually settled in western Canada rather than in his father's homeland of Scotland. In the other instance, mother and son were left to fit themselves back into Japanese society.
27 Barr, *The Deer Cry Pavilion*, 120.
28 Japan Chronicle, *History of Kōbe*, 17.
29 Japan Chronicle, *History of Kōbe*, 17.
30 Ion, *The Cross and the Rising* Sun, II:46.
31 Ion, *The Cross and the Rising* Sun, II:47.
32 Japan Chronicle, *History of Kōbe*, 17.
33 Japan Chronicle, *History of Kōbe*, 17.
34 Japan Chronicle, *History of Kōbe*, 17. See also Kilson. *Mary Jane Forbes Green*, 6.
35 Taylor, *Advocate of Understanding*, 19.
36 NARA, Microfilm 460, Reel 2, U.S. Consular Report 1876, 0094, 0110.
37 Bird, *Unbeaten Tracks in Japan*, II:219–22.
38 NARA, Microfilm 460, Reels 2 & 3, U.S. Consular Reports, Register of American Citizens, 1880, 1881, 1882, 1883.
39 Ishi, *American Women Missionaries*.
40 Powles, *Victorian Missionaries in Meiji Japan*, vol. 4, no. 1 (1987).
41 NARA, Microfilm 460, Reel 4, U.S. Consular Report, 1892, 0714, 5 May 1892, Records the death of the Reverend William James Lambuth.
42 Sugiyama, "Thomas Glover."
43 NARA, Microfilm 460, Reel 1, U.S. Consular Report 1869, 18 December 1869.
44 NARA, Microfilm 460, Reel 1, U.S. Consular Report 1869, 3 June 1869.

45 NARA, Microfilm 460, Reel 1, U.S. Consular Report 1868, 097, and obituary in the *Hiōgo and Ōsaka Herald*, June 1868.
46 Details of the Kirby firm and its principal members are derived from Harold S. Williams Papers, National Library of Australia, Canberra, MS 6681/6.
47 NARA, Microfilm 460, Reel 3, U.S. Consular Report, 1883, 28 December 1883.
48 Mr William Lackie, CBE, provided a notation on Alfred Kirby that appears in the Williams Papers, Box 6.
49 Among those listed as American Consuls or Acting Consuls at Ōsaka and Hiōgo were T. Scott Stewart, appointed 10 February 1868, a confidant of William Seward, Secretary of State, in the Lincoln administration; Major General Julius Stahel, appointed 13 August 1877, a Hungarian emigré who rose to prominence in the Union Army during the Civil War; and Thomas McFadden Patton, appointed 15 May 1884, who had been a legislator in Oregon prior to appointment.
50 Checkland, "The Scots in Meiji Japan," 256.
51 Richard James, Thomas (b. 1827, d. 1900), John Greer (b. 1829, d. 1897), Robert George (1841–1886).
52 Williams Papers, MS 6681/13/5, letter from Thomas Walsh Bartram (grandnephew of the Walsh Brothers) to Williams, 15 May 1963.
53 Williams, *Foreigners in Mikadoland*, 203.
54 Williams, *Foreigners in Mikadoland*, 4.
55 Honjo, *Japan's Early Experience of Contract Management*, 156–8.
56 Williams, *Foreigners in Mikadoland*, 5. Even the great firm of Walsh Hall & Co., like many others, failed to survive. In one of the many depressions, it suffered crippling losses, and the company disposed of its premises in Kōbe and Yokohama to the Hong Kong and Shanghai Banking Corporation, whose address in both ports thereby became "No 2 Settlement." Mitsubishi took over the paper mill. One of the Walsh brothers died and was buried in Kōbe; the other went home to America, and the business folded up, as had many others at an earlier date. As many early merchants discovered, it was often far easier to lose money in Japan than to make it.
57 NARA, Microfilm 460, Reel 2, U.S. Consular Report 1880, 0414, 29 January 1880, Report of Tea Trade for the year ending 31 December 1879.
58 *Hiōgo News*, December 10, 1888.
59 *Retrospective of Kōbe Modernism*, 92.
60 British Foreign Office (hereafter FO), 262 149 #52. Report by Lowder to Foreign Office. The writer of the report offers the opinion that the Chinese,

although restricted in what they were permitted to do by way of trade, acted in a more secretive way to extend these activities.

61 In the twenty-first century, Kōbe continues to have a discernible "Chinatown," which is located more or less in the centre of what was the Native Town or what is now called Nankinmachi Street. The presence of an enduring Chinese enclave is now celebrated by the City, which commemorates the founding of this enclave through a civic designation.

8. Measuring Success in the Concession

1 Harold S. Williams Papers, National Library of Australia, Canberra, MS 6681, Box 8, Kōbe Club; *Hiōgo News*, 18 June 1868; *Hiōgo News*, 10 December 1870.
2 Williams Papers, Box 8.
3 *Hiōgo News*, 27 November 1888.
4 Kanocho, or Kano-cho, is one of the micro-administrative urban units in Kobe. A *cho* recognizes a basic communal grouping connected to a local shrine. The *cho* is part of a hierarchy of units that form the ward or *ku* system of sub-urban administration in Japanese cities. It dates to legislation that was created in 1878. See Ohsugi, *The Large City System of Japan*.
5 Williams Papers, Box 16. See also Noel, *History of Masonry in Japan*.
6 *Hiōgo News*, 10 December 1870.
7 Clark, "The Kōbe Regatta and Athletic Club."
8 *Hiōgo News*, 16 February 1883.
9 *Hiōgo News*, 26 January 1883.
10 *Hiōgo News*, 18 January 1888.
11 Crow, *Highways and Byeways in Japan*.
12 "Tiffin," referring to lunch, is part of the Anglo lexicon derived from India or from other of the colonial bases where trading culture thrived. In this case, the origins are Indian. The word comes from the practice of packing a man's lunch in a tin box to be carried to his place of employment.
13 Williams. *Tales of the Foreign Settlements in Japan*, 227.
14 Murphy, *The American Merchant Experience in Nineteenth Century Japan*, 76.
15 Murphy, *The American Merchant Experience in Nineteenth Century Japan*, 76; Williams, *Tales of the Foreign Settlements in Japan*, 232–4; Abend, *Treaty Ports*, 182–3.
16 Murphy, *The American Merchant Experience in Nineteenth Century Japan*, 76.
17 Murphy, *The American Merchant Experience in Nineteenth Century Japan*, 78.
18 Murphy, *The American Merchant Experience in Nineteenth Century Japan*, 81.
19 As quoted by Barr, *The Deer Cry Pavilion*, 113.

20 Honjo. *Japan's Early Experience of Contract Management*, 30–3.
21 *Kōbe Chronicle*, 19 March 1898
22 *Kōbe Chronicle*, 19 March 1898.
23 *Kōbe Chronicle*, 12 March 1898.
24 *Kōbe Chronicle*, 15 January 1898.
25 *Kōbe Chronicle*, 15 January 1898.
26 *Kōbe Chronicle*, 26 March 1898.
27 *Kōbe Chronicle*, 14 April 1898
28 *Kōbe Chronicle*, 25 January 1898.
29 *Kōbe Chronicle*, 30 April 1898.
30 *British Parliamentary Papers*, " Report 1874"V:502.
31 The Kansai Region encompasses the modern prefectures of Mie, Nara, Wakayama, Kyōto, Ōsaka, Hyōgo, and Shiga. The Kansai region is often compared with the Kantō region, which lies to its east and consists primarily of Tōkyō and the surrounding area. Among the subcultural traits of this region is a pronounced dialect that emanates from the merchant culture of Ōsaka.
32 Matsutani, "Yokohama," 3.
33 Kato, *Yokohama Past and Present*, passim.

Bibliography

Archival Sources

Augustine Heard Collection, Baker Library, Harvard University, Cambridge, MS 766 1835–1892.
British Foreign Office, Correspondence from the British Consul at Hiōgo, Japan, January–June 1868, Public Record Office, Kew, FO 262 148.
British Foreign Office, Letters to Consular Office in Kōbe and Yokohama, Public Record Office, Kew, FO 262 231.
British Foreign Office, General Correspondence, Japan, Public Record Office, Kew, FO 46 158.
Hiōgo Prefecture, Kōbe Siabancho, Correspondence received from the British Consular Office to Governor of Hiōgo, Kobe City Museum.
Hiōgo Court Record, Correspondence directed to the Public Prosecutor 1882–1884, Kōbe City Archives, 57:14.
Japan Gazette Hong List and Directory, Kobe City Archives, 61:1.
Jardine Matheson Archive, Cambridge University Library, Correspondence, Unbound Letters, Kōbe 1868–1876.
The Kenji 1882–1884, Kōbe City Archives 57:14.
Kōbe Foreign Board of Trade, Proceedings, January to July 1869, undated, unsourced clipping, Kōbe City Library, Historical Section.
E.S. Morse diaries, 1877, Phillips Library, Peabody Museum, Salem, Mass., E2, diaries, box #20 b.
Pocket map by Awata, Fukusaburo, titled Shinsen kaisei Heishin shigai no zu: zen [Map of Kobe and Hiogo], published 1880, C.V. Starr East Asian Library, University of California at Berkeley.

National Archives and Records Administration, State Department, United States, Dispatches from the United States Consuls in Ōsaka and Hiōgo 1868–1906, Washington, DC, Microfilm 460, reels 1–6.
Harold S. Williams Papers, National Library of Australia, Canberra, MS 6681.

Newspapers

Hiōgo and Osaka Herald, [published at Kōbe] 1869–75.
Hiōgo News, [published at Kōbe] 1868–88.
Japan Weekly Mail, [published at Yokohama] 1871–96.
The Kōbe Advertiser and Shipping Register, 1879.
Kōbe Chronicle [published at Kōbe] 1897–8.

Published Sources

A Centennial Tribute of Kōbe. Kōbe: City of Kōbe, 1989.
Aiken, S. Robert. *Imperial Belvederes – the Hill Stations of Malaya Kuala Lumpur*. Oxford: Oxford University Press, 1994.
Alcock, Rutherford. *The Capital of the Tycoon*. 2 vols. (London: Spotswood, 1863).
Allen, George C. *A Short Economic History of Modern Japan, 1867–1937*. 2nd rev. ed. London: Unwin, 1972.
Allen, George C., and Audrey G. Donnithorne. *Western Enterprise in Far Eastern Economic Development: China and Japan*. London: Allen and Unwin, 1954.
Allen, Larry. *The Encyclopedia of Money*. Santa Barbara: ABC CLIO, 2009.
An Official Guide to Eastern Asia, Vol. IV, China. Tokyo: Imperial Japanese Government Railways, 1915.
Arima, Sichi. "The Western Influence on Japanese Military Science, Shipbuilding, and Navigation." *Monumenta Nipponica* 19, nos. 3–4 (1964): 352–79. http://dx.doi.org/10.2307/2383177.
"Armenian Merchants of the Seventeenth and Early Eighteenth Centuries: English East India Company Sources." Edited by Vahe Baladouni and Margaret Makepeace. In *Transactions of the American Philosophical Society*, vol. 88, Part 5 1998. http://dx.doi.org/10.2307/1006664.
Barr, Patricia. *The Deer Cry Pavilion: A Story of Westerners in Japan, 1868–1905*. London: Penguin, 1988.
Beasley, W.G. "Self Strengthening and Restoration: Chinese and Japanese Responses to the West in the Mid-nineteenth Century." *Acta Asiatica* 25 (1974): 91–107.

Beasley, W.G. *Great Britain and the Opening of Japan, 1834–1858*. London: Luzac, 1951.
Beasley, W.G. *The Modern History of Japan*. New York and London: Praeger, 1963.
Beauchamp, Edward R. *An American Teacher in Early Meiji Japan*. Honolulu: University of Hawai'i Press, 1976.
Bernier, Bernard. *Capitalisme, société, et culture au Japon, aux origines de l'industrialization*. Montréal: Les Presses de l'Université de Montréal. 1988.
Bird, Isabella L. *Unbeaten Tracks in Japan: An Account of Travels on Horseback in the Interior*. 2 vols. New York: Putnam, 1881.
Black, J.R. *Young Japan: Yokohama and Yedo*. 2 vols. London: Trubner reprints [1881]1968.
Blake, Robert. *Jardine Matheson: Traders of the Far East*. London: Weidenfeld and Nicholson, 1999.
Bowden, Martyn J. "Growth of the Central District of Large Cities." In *The New Urban History: Quantitative Explorations by American Historians*. Edited by Leo. F. Schnore. Princeton: Princeton University Press, 1975.
Bowden, Martyn J. "The Internal Structure of the Colonial Replica City: San Francisco and Others." Unpublished paper presented to the annual conference of the Association of American Geographers, Kansas City, 1972.
Bowden, Martyn J. "The Mercantile City in North America: Theory and Reality." Unpublished paper presented to the annual conference of the Association of American Geographers, San Francisco, 1994.
Bowden, Martyn J. Unpublished field guides to Newport RI, presented to the Eastern Historical Geographers Association Annual Meeting, 1989, and the Bridgetown, Barbados, Eastern Historical Geographers Annual Meeting, 1994.
Bowen, H.V., Margaret Lincoln, and Nigel Rigby, eds. *The Worlds of the East India Company*. Woodbridge: Boydell, 2003.
Boxer, C.R. *The Christian Century in Japan, 1549–1650*. 2nd ed. Berkeley: University of California Press, 1967.
British Parliamentary Papers, Japan Embassy and Consular Reports. Vols. 4, 5, 6, 8, 9, 10. Dublin: Irish University Press, 1971, 1972.
Broadbridge, Seymour. *Industrial Dualism in Japan*. Chicago: Aldine, 1966.
Broeze, Frank, ed. *Brides of the Sea Port Cities of Asia from the 16th–20th Centuries*. Honolulu: University of Hawai'i Press, 1989.
Brunton, Richard Henry. *Building Japan, 1968–1876*. Sandgate: Japan Library, 1991.
Burks, Ardath W. *The Modernizers: Overseas Students, Foreign Employees, and Meiji Japan*. Boulder: Westview, 1985.

Cage, R.A., ed. *The Scots Abroad – Labour, Capital, Enterprise, 1750–1914*. London: Croom Helm, 1985.
Chaudhuri, K.N. *The Trading World of Asia and the English East India Company, 1660–1760*. Cambridge: Cambridge University Press, 1978. http://dx.doi.org/10.1017/CBO9780511563263.
Chaudhuri, K.N. *Trade and Civilization in the Indian Ocean: An Economic History from the Rise of Islam to 1750*. Cambridge: Cambridge University Press, 1985.
Chauduri, Sushil, and Michel Morneau, eds. *Merchants, Companies and Trade Europe and Asia in the Early Modern Era*. Paris: Université de Paris, 1999.
Checkland, Olive. "The Scots in Meiji Japan, 1868–1912." In *The Scots Abroad – Labour, Capital, Enterprise, 1750–1914*. Edited by R.A. Cage. London: Croom Helm, 1985.
Checkland, Olive. *Britain's Encounter with Meiji Japan, 1868–1912*. London: Macmillan, 1989.
Clark, Reg. "The Kōbe Regatta and Athletic Club." *Kansai Timeout* 71 (January 1983): 36–7.
Cooper, Michael. "The Brits in Japan." *Monumenta Nipponica* 47, no. 2 (1992): 265–72. http://dx.doi.org/10.2307/2385239.
Cortazzi, Hugh. "The Pestilently Active Minister: Dr Willis's Comments on Sir Harry Parkes." *Monumenta Nipponica* 39, no. 2 (1984): 147–61. http://dx.doi.org/10.2307/2385014.
Cortazzi, Hugh. *Victorians in Japan: In and Around the Treaty Ports*. London: Athlone, 1987.
Crow, Arthur H., FRGS. *Highways and Byeways in Japan*. London: Sampson Low, 1883.
Crowley, James B. *Modern East Asia: Essays in Interpretation*. New York: Harcourt Brace and World, 1970.
Davenport-Hinds, R.P.T., and Geoffrey Jones. *British Business in Asia since 1860*. Cambridge: Cambridge University Press, 1989.
Davies, Peter. *The Business, Life and Letters of Frederick Cornes: Aspects of the Evolution of Commerce in Modern Japan, 1861–1912*. London: Routledge, 2009.
Elvin, Mark, and G. William Skinner, eds. *The Chinese City between the Two Worlds*. Stanford: Stanford University Press, 1974.
Eng, Robert Y., "The Transformation of a Semi-Colonial Port City: Shanghai, 1843–1941." In *Brides of the Sea – Port Cities of Asia from the 16th–20th Centuries*. Edited by Frank Broeze. Honolulu: University of Hawai'i Press, 1989.
Ennals, Peter. "'Business for Ships Is Miserable Dull': A New Brunswick Mariner Confronts the Waning Days of Sail." *Northern Mariner / Le Marin du Nord* 9 (1999): 23–39.

Fairbank, John K. *Trade and Diplomacy on the China Coast: The Opening of the Treaty Ports, 1842–1854*. Cambridge, MA: Harvard University Press, 1953.
Farrington, Anthony. *The English Factory in Japan, 1613–1623*. 2 vols. London: British Library, 1991.
Finn, Dallas. *Meiji Revisited: The Sites of Victorian Japan*. New York and Tōkyō: Weatherhill, 1995.
Federico, Giovani. *An Economic History of the Silk Industry*. Cambridge: Cambridge University Press, 1997. http://dx.doi.org/10.1017/CBO9780511563034.
Fox, Grace. *Britain and Japan, 1858–1883*. Oxford: Clarendon, 1969.
Frost. Peter. *The Bakumatsu Currency Crisis*. Cambridge: Harvard East Asian Monographs, 1970.
Fujioka, Hiroko. *Kōbe no chushin shiaichi Taimeiho* (off-print 1983).
Gambier, J.W. *Links in My Life on Land and Sea*. London: T. Fisher Unwin, 1910.
Hall, Robert Burnett. "The Road in Old Japan." In *Studies in the History of Culture*. Menasha: American Council of Learned Societies, 1942. 122–55.
Hao, Yen-p'ing. *The Comprador in Nineteenth-Century China: Bridge between East and West*. Cambridge, MA: Harvard University Press, 1970.
Hearn, Chester G. *Tracks in the Sea: Matthew Fontaine Maury and the Mapping of the Oceans*. New York: McGraw-Hill, 2002.
Heubner, Jon W. "Architecture and History in Shanghai's Central District." *Journal of Oriental Studies* 26, no. 2 (1988): 209–69.
Henriques, Robert David Quixano. *Marcus Samuel, First Viscount Bearstead, and Founder of the Shell Transport and Trading Company, 1853–1927*. London: Barrie and Rockcliff, 1960.
Hiōgo Shipping Lists and General Advertiser. Kōbe: 1879.
Hirschmeier, Johannes. "The Japanese Spirit of Enterprise, 1868–1970." *Business History Review* 44, no. 1 (1970): 13–38. http://dx.doi.org/10.2307/3112588.
Hoare, James Edward. "The Japanese Treaty Ports 1868–1899: A Study of the Foreign Settlements." PhD diss., School of Oriental and African Studies, University of London, 1970.
Hoare, J.E. *Japan's Treaty Ports and Foreign Settlements: The Uninvited Guests, 1858–1899*. Meiji Series #1. Folkstone: Curzon Japan Library, 1994.
Hornsby, Stephen J. "Discovering the Mercantile City in South Asia: The Example of Early Nineteenth-Century Calcutta." *Journal of Historical Geography* 23, no. 2 (1997): 135–50. http://dx.doi.org/10.1006/jhge.1996.0046.
Hotham, Edmund G. *Eight Years in Japan, 1873–1881*. London: Keagan Paul and Trench, 1883.

Howe, Christopher. *The Origin of the Japanese Trade Supremacy: Development and Technology in Asia from 1540 to the Pacific War*. London: Hurst, 1996.
Hsia, Ching-lin. *The Status of Shanghai: A Historical Review of the International Settlement*. Shanghai: Kelly and Walsh, 1929.
Huber, J. Richard. "Effect on Prices of Japan's Entry into World Commerce after 1858." *Journal of Political Economy* 79, no. 3 (1971): 614–28. http://dx.doi.org/10.1086/259771.
Hugill, Stan. *Sailortown*. London: Routledge & Kegan Paul, 1967.
Hyōgo Prefecture Exhibitor's Association. *Hyōgo Prefecture and City of Kōbe Panama–Pacific International Exposition*. Kōbe, 1915.
Innes, Robert L. "The Door Ajar: Japan's Foreign Trade in the Seventeenth Century." PhD diss., University of Michigan, 1980.
Ion, A. Hamish. *The Cross and the Rising Sun: The Canadian Protestant Missionary Movement in the Japanese Empire, 1872–1931*. Waterloo: Wilfrid Laurier University Press, 1990.
Ion, A. Hamish. *The Cross and the Rising Sun: The British Protestant Missionary Movement in Japan, Korea, and Taiwan, 1865–1945*. Waterloo: Wilfrid Laurier University Press, 1993.
Ishi, Noriko K. *American Women Missionaries at Kōbe College, 1873–1909*. London: Routledge, 2004.
Itoh, Eiichi. "The Danish Monopoly on Telegraph in Japan – a Case Study in Unequal Communication System in the Far East." *Keio Communication Review* 29 (2007): 85–105.
Jansen, Marius, ed. *Changing Japanese Attitudes toward Modernization*. Princeton: Princeton University Press, 1969.
Jansen, Marius, ed. *The Cambridge History of Japan, Vol. 5, The Nineteenth Century*. Cambridge: Cambridge University Press, 1989.
Jansen, Marius, ed. *The Emergence of Meiji Japan*. Cambridge: Cambridge University Press, 1995. http://dx.doi.org/10.1017/CBO9781139174428.
Japan Chronicle. *History of Kōbe by the Editor and Romance of Kobe by Miss Gertrude Cozad*. Kōbe: Japan Chronicle Jubilee Number, 1918.
Jones, Hazel J. "The Formulation of the Meiji Government Policy Toward the Employment of Foreigners." *Monumenta Nipponica* 23, nos. 1–2 (1968): 9–30. http://dx.doi.org/10.2307/2383106.
Jones, Hazel J. *Live Machines: Hired Foreigners and Meiji Japan*. Vancouver: UBC Press, 1980.
Kato, Yuzo, ed. *Yokohama Past and Present*. Yokohama: Yokohama City University, 1990.
Kazui, Tashiro. *Kinzei Ni-Chō tsūkō bōekishi no kenku* (1981).

Kazui, Tashiro, and Susan D. Videen. "Foreign Relations during the Edo Period: Sakoku Re-examined." *Journal of Japanese Studies* 8, no. 2 (1982): 283–306. http://dx.doi.org/10.2307/132341.
Keeton, George W. *The Development of Extra-Territoriality in China*. 2 vols. New York: Howard Fertig, 1969.
Keith, A.B. *Constitutional History of India, 1600–1936*. London: Methuen, 1936.
Kilson, Marion. *Mary Jane Forbes Greene (1845–1910), Mother of the Japan Mission*. Lewiston: Edward Mellon, 1991.
King, Anthony D. *The Bungalow: The Production of a Global Culture*. London and Boston: Routledge and Kegan Paul, 1984.
King, Frank H.H. *The History of the Hong Kong and Shanghai Banking Corporation, Vol. III, The Hong Kong Bank between the Wars and the Bank Interned, 1919–45; Return to Grandeur*. Cambridge: Cambridge University Press, 1988.
Kiuchi, Shinzo. "Centrifugal and Centripetal Urbanization in Japan." *Proceedings of the IGU Regional Conference in Japan* 1957: 367–71.
Kōbe-shi suido shichyunenshi. [A History of the Kōbe City water works] 1973.
Kornhauser, D.H. *Japan: Geographical Background to Urban-Industrial Development*. London and New York: Longmans, 1976.
Kosambi, Meera, and John E. Brush. "Three Colonial Port Cities in India." *Geographical Review* 78, no. 1 (1988): 126–38. http://dx.doi.org/10.2307/214304.
Kunio, Yoshihara. *Japanese Economic Development: A Short Introduction*. Tōkyō: Oxford University Press, 1979.
Lewis, James B. *Frontier Contact between Choson Korea and Tokugawa Japan*. London: Routledge, 2003.
Li, Lillian M. "Silks by Sea: Trade, Technology, and Enterprise in China and Japan." *Business History Review* 56, no. 2 (1982): 192–217.
Lockwood, Stephen C. *Augustine Heard & Co., 1858–1862: American Merchants in China*. Cambridge, MA: East Asian Research Centre, Harvard University, 1971.
Lockwood, W.W. *The Economic Development of Japan: Growth and Structural Change, 1868–1938*. Princeton: Princeton University Press, 1954.
Ma, Debin. "The Great Silk Exchange: How the World Was Connected and Developed." In *Pacific Centuries: Pacific and Pacific Rim History Since the 16th Century*. Edited by D. Flynn, L. Frost, and A.J.H. Latham. London: Routledge, 1998.
Ma, Debin. "The Modern Silk Road: The Global Raw-Silk Market, 1850–1930." *Journal of Economic History* 56, no. 2 (1996): 330–55. http://dx.doi.org/10.1017/S0022050700016478.

MacMaster, John. *Jardine in Japan 1859–1967*. Groningen, Druk V.R.B, 1967.
Macpherson, W.J. *The Economic Development of Japan c. 1868–1941*. London: Macmillan, 1967.
Mangan, J.A., and James Walvin, eds. *Manliness and Morality: Middle-Class Masculinity in Britain and America, 1800–1940*. Manchester: Manchester University Press, 1987.
Marcovits, Claude. *The Global World of Indian Merchants, 1750–1947*. Cambridge: Cambridge University Press, 2000. http://dx.doi.org/10.1017/CBO9780511497407.
Margrave, Richard D. *The Emigration of Silk Workers from England to the United States in the Nineteenth Century: With Special Reference to Coventry, Macclesfield, Paterson, New Jersey, and South Manchester, Connecticut*. New York: Taylor and Francis, 1986.
Massarella, Derek. *A World Elsewhere: Europe's Encounter with Japan in the Sixteenth and Seventeenth Centuries*. New Haven: Yale University Press, 1990.
Matsui, S. *The History of the Silk Industry in the United States*. New York: Silk Publishing Co., 1927.
Matsutani, Minoru. "Yokohama – City on the Cutting Edge." *Japan Times*, 29 May 2009, 3.
McClain, James L., and Osamu Wakita, eds. *Ōsaka: The Merchant's Capital of Early Modern Japan*. Ithaca: Cornell University Press, 1999.
McGee, T.G. *The Southeast Asian City*. London: Bell, 1967.
McKay, Alexander. *Scottish Samurai Thomas Blake Glover, 1838–1911*. Edinburgh: Canongate, 1993.
Milburn, William. *Oriental Commerce; Containing a Geographical Description of the Principal Places in The East Indies, China, and Japan with Their Produce, Manufactures, and Trade, Including the Coasting or Country Trade from Port to Port; Also the Rise and Progress of the Trade of Various European Nations with the Eastern World Particularly That of the East India Company; from the Discovery of the Passage Round the Cape of Good Hope to the Present Period; with an Account of the Company's Establishments, Revenues, Debts, Assets, &c. At Home and Abroad*. 2 vols. London: Black, Parry & Co., 1813.
Moriyama, Yu. *Kōbe rekishi sampo*. Ōsaka: Sōgensha, 1974.
Morris, L. "Trading with Japan before and after 1858." *Bulletin, Japan Society of London* 54 (February 1968): 2–4.
Morris-Suzuki, Tessa. *A History of Japanese Economic Thought*. New York: Oxford University Press, 1989.
Morse, Edward S. *Japan Day by Day 1877, 1878–79, 1882–83*. 2 vols. Boston: Houghton Mifflin, 1917.

Moulder, Frances V. *Japan, China, and the Modern World; Towards a Reinterpretation of East Asian Development, ca. 1600 to 1918*. Cambridge: Cambridge University Press, 1997.
Murata, Seiji. *Kōbe Kaiko Sanjunenshi*. Kōbe: Kaikō Sanjūnen Kinenkai, 1898.
Murphy, Kevin C. *The American Merchant Experience in Nineteenth Century Japan*. London and New York: Routledge Curzon, 2003.
Murphy, Rhoads. *Shanghai, Key to Modern China*. Cambridge, MA: Harvard University Press, 1953.
Murphy, Rhoads, "Traditionalism and Colonialism: Changing Urban Roles in Asia," *Journal of Asian Studies* 29 (1969): 67–84.
Murphy, Rhoads. *The Treaty Ports and Chinese Modernization, What Went Wrong?* Ann Arbor: Centre for Chinese Studies, University of Michigan, 1970.
Nakane, Chie, and Shinzaburo Oishi, eds. *Tokugawa Japan – the Social and Economic Antecedents of Modern Japan*. Tōkyō: University of Tōkyō Press, 1990.
Nickel and Lyons Ltd. *Yearbook containing customs rates* [and] *general information about the Port of Kōbe*. compiled by J.F. James. Kōbe, ca. 1931.
Nish, Ian, ed. *Britain and Japan Biographical Portraits*. 3 vols. Folkstone: Japan Library, 1994.
Noel, Leo L. *History of Masonry in Japan*. Booklet published in 1963 to commemorate the fifteenth anniversary of the Far East Lodge No. 1, Free & Accepted Masons.
Norman, Egerton Herbert. *Japan's Emergence as a Modern State: Political and Economic Problems of the Meiji Period*. New York: Institute of Pacific Relations, 1940.
Ogborn, Miles. *Indian Ink: Script and Print in the Making of the English East India Company*. Chicago: University of Chicago Press, 2007.
Ogborn, Miles. *Global Lives: Britain and the World, 1550–1800*. Cambridge: Cambridge University Press, 2008.
Ohsugi, Satoru. *The Large City System of Japan*. Tokyo: Papers on the Local Governance System and Its Implementation in Selected Fields in Japan, no. 20, 2011.
Pao, Kuang Yung. "The Comprador: His Position in the Foreign Trade of China." *The Economic Journal* 21, no. 84 (1911): 636–41.
Port of Kōbe. Kōbe: Harbour Department, 1929.
Potts, Francis, and L. Hawks. *A Short History of Shanghai: Being an Account of the Growth and Development of the International Settlement*. Shanghai: Kelly and Walsh, 1928.

Powles, Cyril H. *Victorian Missionaries in Meiji Japan: The Shiba Sect: 1873–1900*. Toronto: University of Toronto–York University Joint Centre on Modern East Asia, vol. 4, no. 1, 1987.

Prakash, Om. *Dutch East India Company and the Economy of Bengal, 1630–1720*. Princeton: Princeton University Press, 1985.

Pringshein, Klaus H. *Neighbours across the Pacific: The Development of Economic and Political Relations between Canada and Japan*. Westport: Greenwood, 1983.

Rawski, Thomas G. "Chinese Dominance of Treaty Port Commerce and Its Implications, 1860–1875." *Explorations in Economic History*, June 1970, 451–73.

Reps, John W. *The Making of Urban America: A History of City Planning in the United States*. Princeton: Princeton University Press, 1965.

Retrospective of Kōbe Modernism. Kōbe: Kōbe City Museum, 1986.

Rowe, William T. *Hankow: Conflict and Community in a Chinese City, 1796–1895*. Stanford: Stanford University Press, 1992.

Sakai, Robert K. "The Satsuma–Ryukyu Trade and the Tokugawa Seclusion Policy." *Journal of Asian Studies* 23, no. 3 (1964): 391–403. http://dx.doi.org/10.2307/2050758.

Sakuda, Yotaro. "In Search of the Origins of the Modern Japanese Economy: History of the Civilization of Great Merchant Families and Zaibatsu." *Senri Ethnological Studies* 26 (1989): 31–49.

Sansom, G.B. *The Western World and Japan: A Study in the Interaction of European and Asiatic Cultures*. Tōkyō and Rutland: Charles Tuttle, 1950.

Sasaki, Seiji. "The Early Development of Kōbe Port." *Kōbe Economic and Business Review* 15 (1968): 17–28.

Sasaki, Seiji. "The Introduction of European-styled Vessels in Japan: A Historical Survey." *Kōbe Economic and Business Review* 10 (1963).

Seidensticker, Edward. *Low City, High City: Tōkyō from Edo to the Earthquake, 1867–1923*. New York: Alfred A. Knopf, 1983.

Shibata, Ginjiro. "Port Labor Conditions in Japan Particularly in Kōbe Port." *Kōbe Economic and Business Review* 6 (1959): 75–90.

Simmons, Duane B., and John Henry Wigmore. *Notes on Land Tenure and Local Institutions in Old Japan*. Ann Arbor: University of Michigan Press, 1979.

Skeldon, C.D. *The Rise of the Merchant Class in Tokugawa Japan, 1600*. Association for Asian Studies, Monograph no. 5, 1958.

Smith, T.C. *Political Change and Industrial Development in Japan: Government Enterprise, 1868–1880*. Stanford: Stanford University Press, 1955.

Society of Map Archivists. *Collected Maps of Japan's Modern Urban Transformation–Ōsaka, Kyōto, Kōbe, Nara, Tōkyō*. Tokyo: Kaiswa Shobo, 1987.

Sorensen, André. *The Making of Urban Japan: Cities and Planning from Edo to the Twenty-First Century.* London: Nissan Institute, Routledge Japan Studies Series, 2002.
Stanislawski, Dan. "The Origin and Spread of the Grid-Pattern Town." *Geographical Review* 36, no 1 (1946): 105–20.
Steeds, David, and Ian Nish. *China, Japan, and 19th-Century Britain.* [Commentaries on British Parliamentary Papers] Dublin: Irish University Press, 1977.
Sugiyama, Shinya. *Bakumatsu and Meiji: Studies in Japan's Economic and Social History.* Edited by Ian Nish. London: International Centre for Economics and Related Disciplines, London School of Economics, 1981.
Sugiyama, Shinya. *Japan's Industrialization in the World Economy, 1859–1899: Export and Overseas Competition.* London: Athlone, 1988.
Sugiyama, Shinya. "Thomas B. Glover: A British Merchant in Japan, 1861–70." *Business History* 26, no. 2 (1984): 115–38. http://dx.doi.org/10.1080/00076798400000023.
Sumiyo, Mikio, and Koji Taira, eds. *An Outline of Japanese Economic History, 1603–1940.* Tōkyō: University of Tōkyō Press, 1979.
Takekoshi, Yosaburo. *The Economic Aspects of the History of the Civilization of Japan.* Vol. III. New York: Macmillan, 1930.
Takano, Fumio. "The City Region Network as Structure of Area." In *Proceedings of the IGU Regional Conference in Japan, 1957.* Tokyo: International Geographical Union, 1959. 486–90.
Tamura, Keiko. *Forever Foreign; Expatriate Lives in Historical Kobe.* Canberra: National Library of Australia, 2007.
Tanabe, K. *Topographic and Historic Factors in Geographic Urban Structure.* Sendai: Institute of Geography, Faculty of Science, Tohoku University, 1975.
Taylor, Sandra C. *Advocate of Understanding; Sidney Gulick and the Search for Peace with Japan.* Kent: Kent State University Press, 1984.
Terry, T. Philip. *Guide to the Japanese Empire.* 3rd ed. Boston and New York: Houghton Mifflin Company, 1928.
Toby, Ronald P. *State and Diplomacy in Early Modern Japan.* Princeton: Princeton University Press, 1984.
Tosh, John. *Manliness and Masculinities in Nineteenth-Century Britain: Essays on Gender, Family, and Empire.* London: Longmans, 2005.
Totman, Conrad. *The Collapse of the Tokugawa Bakufu, 1862–1868.* Honolulu: University of Hawai'i Press, 1980.
Totman, Conrad. *Early Modern Japan.* Berkeley: University of California Press, 1993.

Tsai, Jung-Fang. *Hong Kong in Chinese History: Community and Social Unrest in the British Colony, 1842–1913*. New York: Columbia University Press, 1993.

Tsuru, Shigeto. "Economic Fluctuations in Japan, 1868–1893." *Review of Economics and Statistics* 23, no. 4 (1941): 176–89. http://dx.doi.org/10.2307/1928372.

Vance, James E. *The Continuing City*. Baltimore: Johns Hopkins University Press, 1990.

Vance, James E. *The Merchant's World: The Geography of Wholesaling*. Englewood Cliffs: Prentice Hall, 1970.

Van Dyke, Paul A. *The Canton Trade: Life and Enterprise on the China Coast, 1700–1845*. Hong Kong: Hong Kong University Press, 2006.

Vickers, Robert A., and Jeffery N. Wasserstrom. "Shanghai's 'Dogs and Chinese Not Admitted' Sign: Legend, History, and Contemporary Symbol." *China Quarterly* 142 (June 1995): 444–66. http://dx.doi.org/10.1017/S0305741000035001.

Wallerstein, Immanuel M. *The Modern World System, Vol. I: Capitalist Agriculture and the Origins of the European World Economy in the Sixteenth Century*. New York and London: Academic Press, 1974.

Wallerstein, Immanuel M. *The Modern World System, Vol. II: Mercantilism and the Consolidation of the European World Economy, 1600–1750*. New York: Academic Press, 1980.

Wallerstein, Immanuel M. *The Modern World System, Vol. III: The Second Great Expansion of the Capitalist World Economy, 1730–1840s*. San Diego: Academic Press, 1989.

Waley, Paul, and Nichola Fiévé, eds. *Japanese Capitals in Historical Perspective: Place, Power, and Memory in Kyōto, Edo, and Tokyo*. London: Routledge Curzon, 2003.

Webb, Robert L. "Shipping Light: The Case-Oil Trade to Asia, 1870–1915 and Origins of the Supertanker." *New England Journal of History* 53, no. 1 (Spring 1996): 32–47.

Weigend, Guido G. "Some Elements in the Study of Port Geography." *Geographical Review* 48 (April 1958): 185–200. http://dx.doi.org/10.2307/212130.

Wheatley, Paul, and Thomas See. *From Court to Capital: A Tentative Interpretation of the Origins of the Japanese Urban Tradition*. Chicago: University of Chicago Press, 1978.

Will, John Boxter. *Trading under Sail off Japan 1860–99: The Recollections of Captain John Boxter Will, Sailing Master and Pilot*. Edited by George Alexander Lenser. Tōkyō: Sophia University, 1968 (also Diplomatic Press, Tallahassee, Florida).

Williams, Harold S. *The Foreign Settlements in Japan*. Tōkyō and Rutland: Charles Tuttle, 1958.

Williams, Harold S. *Foreigners in Mikadoland*. Tōkyō and Rutland: Charles Tuttle, 1963.
Williams, Harold S. *Shades of the Past: Indiscreet Tales of Japan*. Tōkyō and Rutland: Charles Tuttle, 1958.
Williams, Harold S. "Shades of the Past: Some Historical Highlights of American Trade in the Kansai in Pre-War Days." *Journal of the American Chamber of Commerce in Japan*, 8 September 1975.
Williams, Jean, and Harold Williams. *West Meets East: The Foreign Experience of Japan*. 2 vols. Rushcutters Bay: Halstead, 1996.
Wittner, David G. *Technology and the Culture of Progress in Meiji Japan*. London: Routledge, 2007.
Wray, William D. *Mitsubishi and the NYK 1870–1914: Business Strategy in the Japanese Shipping Industry*. Cambridge, MA: Harvard University Press, 1984.
Yamaguchi, Keiichiro. "Regional Difference in the Process of Urbanization in Japan." In *Proceedings of the IGU Regional Conference in Japan 1957*. Tokyo: International Geographical Union, 1959. 528–9.
Yamamori, Yumiko. *Japanese Export Furniture with Particular Emphasis on the Meiji Era (1868–1912)*. Web publication accompanying a digital art exhibition of Japanese arts. http://www.euronet.nl/users/artnv/main.html.
Yazaki, Takeo. *Social Change in the City in Japan*. Tōkyō: Japan Publications Trading Co, 1968.
Yip, Christopher. "Neo-Colonial Treaty Port Cities on the China Coast." Paper presented at the symposium "The World Across the University: China in the 1930s and 40s," California Polytechnic State University, San Luis Obispo, 30 May 2009.
Yonemoto, Marcia. "Maps and Metaphors of the 'Small Eastern Sea' in Tokugawa Japan (1603 to 1868)." *Geographical Review* 89, no. 2 (1999): 169–82. http://dx.doi.org/10.2307/216085.
Yoshihara, Kunio. *Japanese Economic Development – a Short Introduction*. Tokyo: Oxford University Press, 1979.

Index

abattoir, 65, 120, 142
Adair, Mr, 72
Adams & Co., 142
Adrian & Co., 39
Alcock, Rutherford, 139, 194
Alt & Co., 112, 123
Amagasaki, 28
American Consul, 8, 52–3, 68, 70, 83, 101, 104, 107, 118, 149, 153–4, 156, 159
American merchants, 11, 93
American Minister, 53. *See also* van Valkenburgh, R.B.; Harris, Townsend
American naval vessels, 23, 74
anti-foreign sentiment, 10
architecture, 19, 75, 127, 131–2, 182
Armenian traders, 3
arson, 55, 65
artisans, 19
Ashton, John, British Consul, 61
Aspinall Cornes & Co., 71
auctions, 34, 38, 197
Augustine Heard & Co., 25, 40, 98, 151, 195
Australia, 10, 40, 71, 97, 134, 136
Avril, P., 51

back streets, 129
Baggally, H.J., 180
Bahama, 72
bakeries, 120, 142, 184
Bakufu, 7–9, 13, 68–9, 74–5, 80, 165, 187
Bangkok, 10
Batavia, 10
bazaar, 135
beach, 140, 166
Bégeux, Monsieur and Madame, 121
Behncke, Ernst, 51
Benne, Mr, 51
Benten no hama, 115, 117, 140
Bernard, C.B., 130, 170
Berry, Dr John (missionary), 160
Bessier, J.N., 51
bilateral treaty, 8
Bird, Isabella, xv–xvi, 138, 159–60, 187
Black Ships, 7, 67
Blackmore, J., 51
Blake, Franklin, 99–101
Blydenburgh, J.S., 46–7
Board & Co., 140
Bombay, 5, 10, 17, 184; Mayor's Courts, 5
Bonger, W.C., 51

230 Index

Bonger Brothers, 132
bordellos, 61, 164
Bovenschen, C., 51
Bowden, Martyn, 19–20, 144–5
Bridges, H.C., 99
British Consul, 34, 52, 55–6, 60–1, 63, 93, 96, 108–9, 115, 126, 149, 151, 162, 183
British East India Company, 4
British Legation, 23, 60
British Minister, 28, 44, 53, 139
British ships, 109
British traders, 5, 8
Browne, Henry St John, 36, 50–1
Brushfield, H.C., 180
Bund, 32, 34–5, 39, 41, 47–9, 51, 64, 99, 120, 122–7, 129, 132, 134, 140, 142, 144, 167, 177, 181, 184, 197
bungalows, 131, 136, 138, 142, 145, 209
business cycles, 86
business practices, 59, 108
business premises, 33, 43, 120, 131, 133, 168

Cabeldu, P.S., 60, 121
Cadiz, 72
Calcutta, 5, 17, 29, 127
Canada, 97, 108–10, 198
Canton, 4–5, 16, 89, 180
Carroll & Co., 140, 142
cattle market, 59–60, 65, 120, 142
Center, Sarah Virginia (missionary), 157
Chamber of Commerce, 49, 63, 68–9, 78–80, 84, 90, 95, 180
Cheefoo, 71
children, 147, 150–1, 155–7, 160, 168, 177, 180
Chile, 57
China, 3–5, 10, 13, 16, 22–3, 26, 30, 32, 42, 55, 64, 69–71, 73, 77, 80, 83, 88, 90, 92, 96–7, 101, 104–5, 108, 132, 147, 151–2, 156, 161–2, 164, 168, 170, 187
Chinatown, 170, 213
Chinese employees, 62, 178, 203
Chinese merchants, 58, 170, 180
Chinese population, 55, 145–6, 148, 150
Chinese raw silk, 90–1
chits, Kōbe Club, 176
cholera, 47, 139, 152–3
Christianity, 3, 6, 157, 160
churches, 19, 144, 151, 157–9, 177
Cocks, Richard, 4
codes of behaviour, 62
commission agents, 19, 144
committee of renters, 51
comprador, 58, 82, 170, 187
consuls, 5, 8, 33, 37, 42–6, 48, 52–3, 57, 59, 61–2, 68, 78–9, 93, 146–7, 158, 167
coolies, 48, 57, 64, 131
Cornes & Co., 34, 39, 93–4, 140
Costa Rica, 72
cotton milling, 183
Council of Foreign Ministers, 44–5
courts, 11, 44, 47, 55, 57, 60, 144, 174
Crutchley Frederick, 48
currency, 67, 69–70, 77, 81–2, 93, 166
Curtis, John, 156
Curtis, W.W. (missionary), 160
customs, 9, 15, 19, 25, 30, 34, 55, 61, 65, 68, 78–9, 112–14, 123, 170, 179. *See also* Eastern Customs House; Western Customs House

daimyō, 22, 187
Davis, Rev. Henry (missionary), 160
Dejima, 4, 6, 182, 191. *See also* Dutch; Nagasaki
Delacamp & Co., 133

diphtheria, 153
disease, 59, 89, 152, 154
Division Street. *See* Nishi-machi
drainage, 26, 46–7, 51
drains, 26, 28, 47–50, 59, 63, 79, 125, 152, 181
Dudley, Julia (missionary), 160
Dutch, 74, 132, 136, 151
Dutch East India Company, 3–4, 6–8, 10, 25, 65
dysentery, 153

Eastern Customs House, 58, 112, 114, 143
Eaton, Mr, 51, 65
elections, 43, 45–6, 53, 146
electric street lights, 124
Elgin Treaty, 8
Emperor Meiji, 9, 192, 199
Eta or Burakumin, 65
Evers, August, 51
export staples, 80, 88

Faber, H., 39, 51
Fairbank, John King, 10
family formation, 145, 147, 150, 155
Favre-Brant, James, 154, 208
females, 62, 92, 150, 155, 160–1
Fire Brigade, 49, 63–4, 173
Fitzgerald and Strome, 140
floods and flooding, 119
Foochow, 16
foreign cemetery, 32, 120
foreign experts, 146
foreign population, 147–50, 154, 158, 170, 183–4
Foreign Powers, 26, 33, 45, 68, 80–1, 139–40, 152
forest, 139, 183
Formosa, 6
Formosa, 72

France, 46, 57, 73–4, 89, 92, 121, 178, 197, 204
franchise. *See* elections
Franco-Prussian War, 96, 172
freehold land title, 43
French, 4, 6, 8, 28, 30, 89, 91, 114, 126, 132, 137, 151, 156–8, 165, 167, 201
Frey, H.J. (shipbuilder), 140
Fukuhara, 61. *See also* bordellos

Gambier, J.W., 25
Gandaubert, Monsieur, 121–2
Gardiner T.W., 72
Gay, Arthur O., 35, 51, 168
German, 30, 105–7, 123, 140, 142, 151, 156, 169, 172, 174, 180, 201, 206
Germany, 57, 73, 178
Gillingham, John, 101
Girls' Home, 160
Glover & Co., 140, 202
Goa, 3
godown, 38–40, 57, 63–4, 99, 103, 112, 134, 140, 142, 154, 197
gold, 69–70, 82–3, 165, 195
gold rush, 178
Goldman, S., 51
Goldsmith, L.R., 208
Goshu, 95
Governor of Hiōgo. *See* Itō, Hirobumi
Greece, 28, 57
Greene, Daniel Crosby (missionary), 158
grid iron plan, 30
grog shops, 25, 36, 55, 61, 124, 151, 164
Groos, J.H., 51, 142
Grosser, E., 51, 142
ground rent, 25, 34, 44, 129, 198
gudang. *See* godown

232   Index

Gulick, Rev. Orramel H. (missionary), 159
Gulick, Rev. Peter (missionary), 159
Gunma, 91

Hakodate, 8–9, 182–3, 192–3
Hall, Frank, 168
Hall, George, 168
Hankow, 30, 32, 101
Hansell, Alexander (architect), 132
harbour, 4, 13–15, 17, 22–3, 25, 28, 30, 33, 38, 41, 47, 73, 112–13, 116, 126, 135, 140, 167, 183–4, 188, 195
Harris, Townsend, 7, 168
Hart, J.W., 28, 46–9, 61, 132, 182, 196
Hassage, Mr, 72
Hawaii, 71, 151, 159
Heard & Co., 98–9, 101, 151, 197
Hecht Lilenthal & Co., 142
Heinemann, Paul, 46, 197
Herhausen, Otto, 51
Hill, 37–8, 52, 111, 116–17, 127, 129, 136–9, 145, 149, 173, 178, 181–2
Hiōgo Gas Co., 170
Hiōgo Hotel, 122, 135, 142
Hiōgo News, 38, 47, 55, 57–8, 61, 73, 97, 105, 112, 115, 119, 124–5, 132, 138, 142, 152–3, 173
Hiōgo and Ōsaka Herald, 45, 54–5, 58, 65–6, 68, 96
Hiōgo Prefectural Board of Health, 153
Hirado, 4
Hiroshima, 16, 118
Hoare, James, 147
Hobson & Co., 142
Holiday and Wise & Co., 94
Holme, R., 51
Hong Kong, 4, 16, 20, 25, 71, 83, 98, 121, 151, 156, 162, 166

hongs, 123, 151, 175, 178, 182, 187
Honjo, Yuki Allyson, 11
Honshū, 9
Hornsby, Stephen, 17
Horton, H.H., 72
hospitals, 32, 49, 54, 65–6, 120, 152–3, 200–1
hostels, 19, 36
hotels, 19, 37, 41, 121–2, 129, 132, 144, 164, 172, 175
Hozu, 16
Huangpu River, 196
Hunnink, Dr Schokker, 39
Hunt, H.J., 123
Hunt & Co., 123
Hunter, E.H., 36

Ida, 95
Ikuta River, 23, 32, 49, 61, 65, 119–20, 127, 142–3, 166, 173
Ikuta Shrine, 137
Imperial Railroad, 116–18
India, 3–6, 20, 23, 27–8, 30, 88, 97, 108, 132, 151, 168
Inland Sea. *See* Seto Sea
insurance, 79, 144, 170, 201
Itō, Hirobumi, Governor of Hiōgo, 27–8, 32, 49, 58, 61, 79, 195–6
Iveson, H., 51
Iwasaki, Yatarō, 77, 169

jails, 50, 54, 56, 62, 144
Japanese Department of Finance, 143
Japanese houses, 25, 140
Japanese land tax, 45
Japanese legal proceedings, 43, 47
Japanese Normal School, 124
Japanese press, 11
Japanese shipping, 74, 76
Jeddah, 10

Jenks, DeWitt (missionary), 160
Jewish traders, 3
Joseph, Mr, 51

Kaga, 102
Kaga Foundry, 140
Kaigan-dori. *See* Bund
Kamo River, 16
Kanagawa. *See* Yokohama
Katsura River, 16
Kencho, 181
Kipling, Rudyard, xvi, 121, 206
Kirby, E.C., 36, 156, 166
Kirby & Co., 61, 142–3, 165
Kisaburo, 47–8
Kitano-cho, 127, 137
Kōbe, 16, 72, 80, 110, 150; arrival of traders, 36; churches, 158; coastal foreland, 16; commencement of trade, 85; growth after 1900, 184; harbour, 15; initial site assessment, 26; land price, 33–4; plan of, 30; preservation of older houses, 129; railway, 118; rice mill, 124; rivalry with Yokohama, 85; sailor town, 164; shipping services, 71–3; silk export, 80, 93; social class, 163; trading hinterland, 16; tree planting, 126
Kōbe City Museum, 129–30
Kōbe Club, 172, 173
Kōbe College, 160
Kōbe Electric Light Co., 125
Kōbe Ice Co., 170
Kōbe Iron Works, 166
Kōbe Paper Factory, 143
Kōbe Pier, 65, 113–15
Kochi, 3
Korthalls W.C., 51
Kurile Islands, 7
Kyo-machi, 33, 41, 120–1, 129, 172–3

Kyōto, 9, 16, 22–4, 80, 94, 96, 102, 111, 115, 117–19, 149, 159, 169, 171, 178, 181, 183

Lake Biwa, 16, 94, 118, 156
Lambuth, Rev. James William (missionary), 161
Lambuth, Rev. Walter Russell (missionary), 161
land auction, 32–4, 44
land leases, 44, 52
land lottery, 36
land rent, 100, 144
land speculation, 117
land tax, 50, 52
landlord and tenant relationships, 37
leisure activities, 172. *See also* Recreation Ground
Lester, Henry (architect), 127
levees. *See* Ikuta River
lighthouses, 146
Lippert, W.E., 132
livery keepers, 164
Look, Mr, 51
lot sizes, 30
lotteries, 36
Lowden, Edward, 34, 55, 197
Lyons, J., 89, 72

Macao, 3
Macclesfield, 89, 94
Mackenzie, Kenneth Ross, 46, 167
Madras, 5, 10, 17, 29, 127
Malacca, 3, 197
Malay policemen, 56
Malaya, 197
Mancini, N., 142
mariners, 36, 65, 115, 152
Marmalestein, Mr, 51
Masonic Lodges, 158, 173–4
Mathie, Mr, 72

Maye-machi, 127
McCailin, Mr, 72
mechanics and blacksmiths, 36
Mechanic's Institute, 50, 198
Meiji Government, 9, 45, 75, 116, 182, 195–6
mercantile activity, 15, 17, 19–20, 67, 70–2, 79, 80, 106, 112, 145, 147, 167, 181
mercantile cities, 17, 19, 129, 145
mercantile ports, 20, 32, 111, 145, 171
mercantile triangle, 143
merchant vessels, 23, 71–3
merchants, 17, 67, 69, 74–7, 100, 112, 114, 179, 183
Merikan Hatoba, 112, 140, 168
Messageries Impériale, 89, 202
Messrs Reynell & Co., 64
Mexican dollar. *See* currency
mission societies, 157–9
missionaries, 146, 157, 160–1, 170
missionaries: American, 138, 152, 157, 159; Episcopalian, 158; female, 160–1; Great Awakening, 160; medical, 160; Portuguese, 6; wives of, 177
Mitsubishi, 76–7, 169, 202
Mombasa, 3
morphology of cities, 13, 17
mortality rates, 147, 153
Moto-machi Street, 125, 132, 135, 158, 208
Mourilyan, Heiman & Co., 51, 61, 105
Municipal Building, 49–50, 54, 56, 64, 134, 144–5
Municipal Council, 28, 34, 42–9, 53–4, 58–9, 61–4, 78, 118, 125–6, 144, 146, 154, 170, 173–4, 181

Municipal Hall, 49, 51, 62, 121, 127, 181
municipal taxes, 34, 38, 52–3, 138, 174
municipal water system, 125
Murphy, Rhoads, 10
Muslim merchants, 3
Myburgh, Francis G., 23, 25, 34, 66, 196

Nagasaki, 4, 6, 8–10, 25, 27, 30, 35–6, 71–4, 83, 98, 123, 133–5, 137, 152, 162, 165, 167–8, 182, 195, 197
Nagoya, 16, 102, 117, 118
Nakashima, 48
Nanban, 3
Naniwa-machi, 121, 137
Nara, 16, 118
Native Bund, 122, 124, 135, 164
Native Town, 25, 30, 32, 34–8, 40, 47, 49, 56, 59, 61, 63–4, 71, 79, 98–100, 105, 111–13, 116–17, 123–6, 129, 131, 135, 140, 146, 149, 152, 155, 158, 164, 168, 170, 172, 175, 194, 197
naval vessels, 23, 74
Netherlands Trading Society, 39
New England, 131, 159
New York, 35, 72, 107–8, 157, 168
New York, 72
Nichi Nichi Shinbun, 52
Niigata, 8–9, 182–3, 193
Nishi-machi, 112, 123–4, 127, 135, 140, 181
Nishinomiya, 28

Ogborn, Miles, 191
Ōkayama, 16, 118, 149
opium, 4, 42, 106, 162
Oregonian, 72
Ōsaka, 79, 95, 199, 202

Osaka: Chinese merchants, 55, 180; cholera, 153; connections to Kyōto, 23; cotton textiles, 87; currency valuation, 69; financiers, 184; growth pole, 182; harbour, 15, 23, 183; hinterland, 16, 102; industrial centre, 22, 111; manufacturing, 9; missionaries, 157, 160; Mita Paper Mill, 169; opening of treaty port, 9, 23; population, 15, 17, 149, 183; port development, 111, 113; railway, 100, 117–18, 137; road connections, 115; sale of lots, 99; silk, 80, 95–6; telegraph, 115; textile manufacturing, 94; trade, 15, 49, 70, 79, 81, 178; travel to, 171; unrest at, 74; wooden match industry, 183
Ōtsu, 16, 96, 118, 156, 203

Pacific Mail Steamship Co., 35, 72, 75, 77, 157
paper making, 169
Papier Fabrik. *See* Umezu Co.
Parks, Sir Harry, 23
passports, 118–19, 138
Peninsular and Oriental Steam Navigation Co., 34, 72
Perry, Commodore Matthew, 6–8, 67, 116, 157, 182
Peru, 27, 57
Phelps, Alfred, 71–2
Poland, 57
police force, 45, 54–6, 62–3, 181
population: Chinese, 54–5; female, 62; foreign, 147–50, 154, 157–8, 170; growth, 149; impermanent, 163; Japanese, 153, 157; mariners, 57, 65; native, 149; women and children, 150

Port Island, 184
ports, 8, 10, 12, 14, 16, 25, 81. *See also* treaty ports
ports: American, 109; characteristics, 14–15; India, 20; Japanese, 11, 75; morphology, 14, 30; New York, 105
Portuguese, 3–4, 6, 10, 127, 151, 191
Presidency cities, 5, 17
promenade, 32, 41, 126–7, 135, 176
Prussia, 46, 52, 197
Public Garden, 30, 126, 134
public hygiene, 26, 51, 125, 152–3
public works, 44–5, 53–4, 113, 115

Qing dynasty, 4

Raffles, Sir Thomas Stamford, 7
railways, 115–18
Rangoon, 10
real estate advertising, 131
Recreation Ground, 113, 120, 127, 143, 173–4, 181
retail shops, 19, 41, 121, 144
Reymond, Monsieur, 121–2
rice, 16, 22–3, 47, 52, 82, 84, 198
rice mill, 124
Richter, Mr, 51
Richter and Reinhardt, 142
Rōkkō Island, 184
Rōkkō Mountains, 22–3, 119, 127, 139
rōnin, 55, 188, 199
Russell & Co., 151
Russia, 183
Russians, 4, 8, 151
Ryūkū Islands, 13

Saigoku High Road, 22
Sakaye-machi, 124–5
samurai, 28, 188, 193

San Francisco, 90, 108–9, 155, 164, 166
Sannomiya, 113, 117, 143, 155, 169
Sanyo rail line, 118
Satsuma han, 9, 75, 77, 165
Satsuma-Ryūkū trade, 194
Scandinavia, 57
Schultze Reis & Co., 39
Schut, J., 51
Scott, J., 51
sea wall, 26, 41, 51, 135
seamen, 54–7, 144, 147, 149, 153–5, 163–4
Semiche Faber & Co., 39
sendōs, 57
Seto Sea, 9, 111
Shanghai, 5, 10, 16, 20, 25, 30, 32, 36, 42, 47, 57, 71–3, 77, 83, 89, 94, 98, 104, 121, 126–7, 131, 151, 162, 164–8, 193, 195–6, 200–1
Shikoku, 16, 22, 102
Shimo Yamate-dori, 136
Shimoda, 8–9, 182–3, 192
ship chandlers, 36
Shōgunate, 4, 187, 195–6
silk, 4, 16, 80, 89, 91
silk: decline of trade at Kōbe, 96; exports from Japan, 89–90; factory production, 89; France and Italy, 89, 91; grades, 80; inspection house, 91; Japanese industry, 90–1; Maebashi, 91; Marseille and Lyon, 89; prices, 93; production problems in Europe, 80; sericulture, 88; silk exports, 96; silk merchants, 92; silkworm, 88; trade at Kōbe, 88, 95; United States mills, 89; weaving, 89; Yokohama, 80, 96
Simon, Mr, 51

Singapore, 10, 127
Sino-Japanese War, 148, 180
Skipworth & Hammond General Store, 121
smallpox, 65, 152, 153
Smith Baker & Co., 35, 39, 71, 105, 123, 134, 152, 157
smuggling, 42, 69
social clubs, 19, 144, 175
sports, 155, 173, 176
St Francis Xavier, 3
Stahel, Major General Julius, 62, 105, 107
Storey & Smedley, 132
Storey, W., 72
street lighting, 49, 170
Suez Canal, 73, 107, 162, 202
Suwa-yama, 137

T. Eaton Co., 206
Talcott, Eliza (missionary), 160
tarriffs and duties, 68
taxes, 37, 52, 54
tea, 4, 16–17, 33, 63, 79–80, 84, 88, 93, 97–112, 114, 118, 120–1, 129, 132, 152, 156, 160, 168, 170, 176, 178, 184
Tempo famine, 7
Textor & Co., 36, 39, 140
Thayer & Co., 140
theft, 39, 58, 61
tobacco, 81
Tokugawa, 4, 6–7, 12–13, 22, 66, 74, 182, 187, 192, 200
Tōkyō, 9, 23–4, 44, 69, 86, 96, 115, 117–18, 120, 156–7, 160, 169, 183–4, 214
Tōkyō Bay, 7
Tosa, 75, 77, 102
town plan, 28, 51

Townley, Frederick, 72
trade goods, 14–16, 38, 68–9, 118, 162
treaty ports, 11, 23, 32, 43, 71, 89, 127, 146; Chinese, 26, 30, 55, 127, 170; Hiōgo, 23, 55; Japanese, 11–12, 75, 182; Nagasaki, 71, 161, 182; Shanghai, 57; Yokohama, 25, 71, 161, 183
Trotzig, Herman, 53, 181
typhus, 47, 153

US Consulate, 49
Uji River, 140
Umezu Co., 169
United States, 46, 57, 62, 70, 73–4, 90, 97–8, 105, 107, 109–10, 160–1, 164, 168–9, 178, 184, 197–8, 201
urban planning, 26

Van der Vot, Mr, 72
van Valkenburgh, R.B., 23
Vancouver, 90, 109
Vandervlies, Mr, 51
Vereenigde Oost Indische Companie. *See* Dutch East India Company
voters, 54
Vulcan Foundry, 140

Wachtel Groos & Co., 142
Wainwright & Co., 138, 142, 164–5

Wallerstein, Immanuel, 191
Walsh Hall & Co., 142, 167–9
Wanxian, 5
warehouses, 14, 19, 23, 30, 40, 78, 144
Warren, Charles Frederick, 51, 158
Warren Tilson & Co., 39
Water Street, 142
water taxis, 49
waterfront district, 19
Waters, Mr, 51
Western Customs House, 71, 100, 113, 140
whaling, 7, 74
wharfs, 14, 19, 113
Wignall Ship Yard, 140
Williams, Harold S., 12, 168
wives, 138, 147, 155, 160, 177

Yamato River, 16
Yangtze River, 4, 30
Yedo, 9, 22, 64, 69, 94, 116, 182–3, 192, 195. *See also* Tōkyō
Yodo River, 16
Yokohama, 8–12, 16, 20, 25–7, 30, 32–6, 42–5, 62, 70–3, 80, 83, 85–6, 88, 90–8, 102, 107, 109, 115, 117–18, 123, 151, 155, 162–6, 168–9, 178, 183–4, 193, 195, 197, 210, 212

Japan and Global Society

Yoshihide Soeya, Masayuki Tadokoro, and David A. Welch, eds., *Japan as a "Normal Country"? A Nation in Search of Its Place in the World*

Leonard J. Schoppa, ed., *The Evolution of Japan's Party System: Politics and Policy in an Era of Institutional Change*

Masato Kimura and Tosh Minohara, eds., *Tumultuous Decade: Empire, Society, and Diplomacy in 1930s Japan*

Tomoko T. Okagaki, *The Logic of Conformity: Japan's Entry into International Society*

Peter Ennals, *Opening a Window to the West: The Foreign Concession at Kōbe, Japan, 1868–1899*

www.ingramcontent.com/pod-product-compliance
Lightning Source LLC
Chambersburg PA
CBHW020402080526
44584CB00014B/1140